D1084464

The Infested Mind

The Infested Mind

Why Humans Fear, Loathe, and Love Insects

Jeffrey A. Lockwood

OXFORD UNIVERSITY PRESS

OXFORD
UNIVERSITY PRESS

Oxford University Press is a department of the University of Oxford.
It furthers the University's objective of excellence in research, scholarship,
and education by publishing worldwide.

Oxford New York
Auckland Cape Town Dar es Salaam Hong Kong Karachi
Kuala Lumpur Madrid Melbourne Mexico City Nairobi
New Delhi Shanghai Taipei Toronto

With offices in
Argentina Austria Brazil Chile Czech Republic France Greece
Guatemala Hungary Italy Japan Poland Portugal Singapore
South Korea Switzerland Thailand Turkey Ukraine Vietnam

Published in the United States of America by
Oxford University Press
198 Madison Avenue, New York, NY 10016

Library of Congress Cataloging-in-Publication Data
Lockwood, Jeffrey Alan, 1960–
The infested mind : why humans fear, loathe, and love insects / Jeffrey Lockwood.
p. cm
Includes index.
ISBN 978-0-19-993019-7 (alk. paper)
1. Insect phobia. 2. Fear. I. Title.
RC552.A48L63 2013
616.85'225—dc23
2013005383
9780199930197

9 8 7 6 5 4 3 2 1
Printed in the United States of America
on acid-free paper

To my dear friends who struggle with mental illnesses of various forms and severities: CA, AC, TC, RC, RE, SG, CJ, CL, LT, MV, and KV—may there be a time when the social stigma is lifted and you can be treated with the compassion and care you deserve rather than being burdened with shame and anonymity.

"What sort of insects do you rejoice in, where you come from?" the Gnat inquired.

"I don't rejoice in insects at all," Alice explained, "because I'm rather afraid of them."

—Lewis Carroll, *Through the Looking Glass* (1872)

CONTENTS

LIST OF ILLUSTRATIONS

ACKNOWLEDGMENTS

A wide-ranging project concerning the psychological relationship between humans and insects requires creative and thoughtful research assistants, and I am grateful for the determined work that was provided by Bob Weatherford, Ryan Ikeda, and Sarah Hartsfield. Given vague and open-ended instructions to "Find something about _____" or "See where this leads . . . ," they displayed diligence, intelligence, and curiosity that often revealed stories and information that I could not have imagined.

As family, friends, and colleagues learned of this project, they were most helpful in offering ideas, articles, and truly bizarre tidbits from some rather remarkable sources. For these oftentimes revealing and sometimes disturbing contributions, I thank Ken Gerow, Susanna Goodin, Franz-Peter Griesmaier, Erin Lockwood, Ethan Lockwood, and Scott Shaw. Rob Colter indulged me in my quest for Greek terms to make concepts appear erudite and allowed me to bastardize the language of his beloved academic field, ancient philosophy. And Joe Ulatowski provided wonderfully incisive and constructive suggestions regarding many elements of the book. Most important, my wife Nancy provided constant support for my writing.

At the University of Wyoming, I received encouragement for this project from my "bosses": Susanna Goodin, head of the department of philosophy, and Beth Loffreda, director of the master of fine arts program in creative writing. A sabbatical leave granted by the university provided valuable time for me to work on the book. And at Oxford University Press, I received the able encouragement and support of my editor, Jeremy Lewis, as well as Tisse Takagi, who has since moved on to another publishing house. The index was provided through the consummate skills of Margery Niblock.

I am deeply grateful for the time, patience, and thoughtfulness of those who granted me interviews. Dick Nunamaker and Heather Story provided compelling views into their work, lives, and passions. Will Robinson and Scott Schell offered powerful stories of delusory parasitosis (concerning people who came to these entomologists, not Will and Scott themselves). Harvey Lemelin set me straight regarding the shortcomings of biophilia. Tony Earls provided vivid and forthright views into the world of bed bugs from the perspective of a university director of residence life. And Elizabeth Rasmussen was a wellspring of information about Morgellons syndrome.

Brett Deacon, the director of the Anxiety Disorders Research Laboratory at the University of Wyoming, was extraordinarily generous with his time and expertise throughout this project. He provided a careful, scientific review of several chapters in the book, and his expertise in this regard was invaluable. However, any and all errors concerning psychological concepts—as well as physiological, anthropological, sociological, historical, and literary/cinematic claims—are most assuredly my own.

PROLOGUE

THE INFESTATION BEGINS

I stared warily across the barbed-wire fence as the dust kicked up by my truck hung over the road. The view across Mr. Martin's sun-baked pasture was eerie. The sagebrush, normally gray-green with leathery leaves, were skeletons. The yuccas were shredded, as if they had been attacked by a demented rancher armed with a lawn trimmer. Even the Canadian thistles looked like refugees from a devastating hailstorm, except it hadn't rained in nearly a month. The grass was baked to a golden crisp and cropped to a height of a couple of inches, as if that crazed rancher had also owned a riding mower.

Grasshoppers were clinging to the skeletons of the sagebrush and blanketing the shady sides of the fence posts to avoid the searing heat of the soil. When I took a few steps into the field, they exploded from the sparse vegetation. The density of life was dizzying. Incredulous, I continued into the field, heading toward a gully that promised an encounter beyond anything I had experienced in a decade of work on the Wyoming prairies.

* * *

Until that day, Whalen Canyon had never been a disturbing place. This humble feature of the southeastern Wyoming steppe does not have the vertiginous quality of the Grand Canyon or the bear-driven anxiety that comes with hiking in Glacier National Park. Whalen Canyon is not Sedona, Ayers Rock, or Lhasa, where people's lives are transformed. So I wasn't looking for horror, epiphany, or change of any kind. I was there to gather ecological data and was utterly unprepared for what happened on that sere expanse of grassland.

Whalen Canyon is not much of a canyon, at least where I had my encounter. Rather, it is more like a mile-wide expanse of native grasses sloping gently from rocky hills down to the Platte River. The canyon—more of a cleft between rocky hills—is the natural feature closest to our research plots, so that's what we called the study site. The road out of Guernsey, a rural community of a thousand or so windblown souls known for its National Guard camp and the

Oregon Trail wheel ruts, leads to the canyon. But what normally brought people to the town was not what attracted me and my research team from the University of Wyoming. Rather, we were drawn to some of the most dependably prolific grasshoppers in the West.

For an entomologist dedicated to understanding the ecology of rangeland grasshoppers and developing better ways of suppressing their outbreaks, Whalen Canyon is a godsend. Even in "lean years" (meaning few grasshoppers but plenty of grass for ranchers' cattle), we could usually count on populations of at least ten grasshoppers per square yard in this area. Such a density typically would not justify the cost of an insecticide treatment, but the numbers were sufficient for testing various insecticides. In "good years," Whalen Canyon could produce phenomenal numbers. And 1998 was producing a bumper crop.

So it was that on a bone-dry day in early July, I stopped by some small plots in which we were testing a new insecticide. I had made the trip by myself, as my research crew was working on another project. Within the plots, the grasshopper numbers were running just five or six per square yard, down nearly 80 percent in the two weeks since we had treated the plots with an experimental compound. Taking on industry contracts to test products was one of my least favorite endeavors, but these ventures paid handsomely and provided funds that I could parlay into less lucrative but far more interesting ecological research.

When the guys had checked the plots a week earlier, they had told me that to the north, where the road swung toward the mouth of the canyon and deep draws were etched into the prairie, the grasshoppers were reaching biblical proportions. I had encountered some high densities before. Working with forty or fifty grasshoppers per square yard was oddly thrilling, and I wanted to see this infestation for myself. I have to wonder now how my life might have been different if I had decided instead to head back to Laramie and take care of the backlog of mail that had accumulated during the field season. But it was like coming across a horrible accident along the highway—once you've stopped to see, there's no erasing the memory.

The earthen banks rose above my head as I descended into a draw. In the gulch, where only a hint of green vegetation remained, the grasshoppers had amassed into a bristling carpet of wings and legs. My arrival incited a riot, the carpet irrupting into seething chaos. Rather than waves of movement parting in my path, there was sheer pandemonium. Grasshoppers boiled in every direction, ricocheting off my face and chest. Some latched onto my bare arms and a few tangled their spiny legs into my hair. Others began to crawl into my clothing—beneath my shorts, under my collar. They worked their way into the gaps between shirt buttons, prickling my chest, sliding down my sweaty torso. For the first time in my life as an entomologist, I panicked.

A Swarm of Locusts falling upon and devouring a Wheat-field.

Figure 0.1
This nightmarish woodcut comes from C. V. Riley's *The Locust Plague in the United States* (1877). The species that devastated the pioneers was closely related to the species that I encountered more than a century later in southeastern Wyoming—and the sense of suffocating numbers, chaotic movement, and overwhelming presence conveyed by this image is frighteningly familiar, perhaps more so than any photograph can capture.

* * *

I was a child the last time I felt the rising terror of losing myself, engulfed within a suffocating amorphous presence. In my youthful nightmare, a visceral panic rooted in a primal horror would sweep through me. Like a swelling globule of mucus or fat, the protean mass was utterly indifferent to me as it inexorably filled the room. The inescapable, bloating presence became a recurring visitor, and I'd wake up twisted in my sheets with my heart pounding. I dreaded falling asleep again following one of its smothering visits.

The dread lasted until I found a method to control these episodes, not by suppressing the feeling of oppressive enormity but by inducing it on my own terms. In adolescence, I could use lucid dreaming to gradually evoke a vivid felt-sense of being an infinitesimal mote in infinite space. Experiencing the disappearance of myself in this manner was enchanting—and terrifying.

As an adult, I either lost or forgot this strange practice. The only echoes were a vague discomfort in crowds, an intense reaction to Hitchcock's *The Birds*, and an irrational fascination with the concept of infinity. Most childhood nightmares fade with time. Mine metamorphosed on the Wyoming prairie.

I remember moments: frantically sweeping the grasshoppers from my clothing, shaking them from my hair, and somehow making it back to the truck. I can clearly recall only the blind, irrational, unaccountable terror. Shaking, I climbed inside the truck and slowed my breathing. Then I began the two-hour drive home. The smallest shift of every grass seed lodged in my clothing and each barely detectable tickle from a drop of sweat evoked an anxious slap or hurried brush, as if the grasshoppers were still clinging to me. As these sensations faded over the miles, I tried to forget what had happened. But I couldn't.

I was, after all, an entomologist. I had encountered insects in all sorts of contexts for many years. And this experience was like being an established surgeon who one day fainted at the sight of blood. Only worse. A surgeon is supposed to be empathetic. It was more like being a riveter on a skyscraper who suddenly experiences a heart-pounding dread of heights. I had lost my nerve. As days passed, I began to wonder if the panic would return unexpectedly, when I could not so easily hide my humiliating reaction.

I had lost my ability to engage insects dispassionately. They had worked their way into my psyche, and to a troubling extent my research became personal. It's not that I sought to destroy the source of my anxiety; I did not begin to take pleasure in killing grasshoppers as if they were now my nemeses. If anything, quite the opposite. These creatures became deeply affecting to me—they were able to enchant my imagination, not merely engage my cognition. A scientist ought to be objective—and I no longer was.

* * *

The experience of being buried alive by life in that rangeland draw challenged my sense of psychological well-being. Let's just call it what it is—my mental health. It changed me in ways that would haunt me for years, and in retrospect it may have catalyzed my eventual move from the sciences to the arts and humanities (I now teach in the university's department of philosophy and creative writing program). Ironically, the memory loosened its grip when I returned to science, this time psychology rather than entomology. I needed to make sense of that day and the network of experiences that led up to and followed from it.

This book was initially motivated by my need to understand myself, to reassert control. The results of my exploration might have been therapeutic but not terribly interesting to others, except in providing a view into the mind of a disturbed scientist. However, my experience—or some version of it—is common in contemporary human society. About one person in ten develops a phobia in the course of his or her life.[1] Fears of animals and heights are most common, but nearly fifty million people experience anxiety involving animals, and eleven million people wrestle with entomophobia.[2]

Recent stories of how people perceive the bed bug explosion and the irrational responses to these interlopers cast a bright light on this shared anxiety about insects. But our emotional response to insects on our bodies and in our homes is not merely a modern, socially constructed phenomenon. Rather, it is a vital part of being human. Our perception of insects is deeply rooted in our species' evolutionary past. And so if insects and their kin evoke in you a moment's hesitation, a persistent shudder, or even a curious enchantment, then I invite you to join this expedition into the infested mind.

NOTES

1. Aaron T. Beck, Gary Emery, and Ruth L. Greenberg, *Anxiety Disorders and Phobias: A Cognitive Perspective* (New York: Basic Books, 2005), chap. 6; David H. Barlow, *Anxiety and Its Disorders: The Nature and Treatment of Anxiety and Panic*, 2nd ed. (New York: Guilford Press, 2002), chap. 1.
2. Martin M. Antony and David H. Barlow, "Specific Phobias," in Barlow, *Anxiety and Its Disorders*; William W. Eaton, Amy Dryman, and Myrna M. Weissman, "Panic and Phobia," in *Psychiatric Disorders in America: The Epidemiologic Catchment Area Study*, ed. Lee N. Robins and Darrel A. Regier (New York: Free Press, 1991).

The Infested Mind

CHAPTER 1

The Nature of Fear—and the Fear of Nature

My experience of being overwhelmed by grasshoppers is difficult for me to recount precisely, but I'm not alone in this struggle. In talking to dozens of individuals about their unsettling encounters with insects, I have found that what emerges is a muddle of emotions that are variously described as terror, panic, and revulsion—often followed by a sense of confusion, humiliation, and anxiety. It turns out that people are not very good at precisely naming their emotions, both because we rarely experience just a single emotion at any given time and because even individual emotions are difficult to characterize.

Dozens of emotions arise and interlace over the course of a lifetime, from affection and boredom to shame and worry. Different cultures exhibit idiosyncratic emotions, or at least use unique terms for particular feelings. For example, *aviman* is the Indian term for prideful, loving anger and *oime* is the feeling of dutiful indebtedness in Japanese culture.[1] Americans have no specific word for either of these.

The study of emotions might be a hopeless tangle if psychologists had not sought and found common ground across our species. People of all cultures, races, genders, and ages experience a set of six emotions: happiness, sadness, surprise, fear, disgust, and anger (some psychologists also include trust and anticipation). Nor are most of these emotions unique to humans, being evident in our primate brethren and a variety of other animals. From these building blocks, we construct more elaborate emotions, such as disappointment (sadness plus surprise) and contempt (disgust plus anger).[2] Of particular interest to those who recoil from insects is the feeling of horror, which has been characterized as disgust-imbued fear.

So let's begin with fear and its emotional allies (we'll get to disgust in due course). For many, fear feels like the core emotion of an aversive encounter with an insect—or a few million insects.

Figure 1.1
Facial expressions convey universal emotional states in humans—and in our primate relatives. An observer has no difficulty in interpreting the psychological distress expressed in the shape of the mouth and the eyes (images by JelleS and Lorenzo Sernicola through Creative Commons).

THE EMOTIONAL ANTE: ANXIETY AND FEAR

Being scared is no simple matter. Consider the range of words that we use to describe this feeling: anxiety, apprehension, dread, fear, misgiving, panic, terror, trepidation, and unease. Although our language allows for many nuances, psychologists have distilled these expressions of our mental experiences into two foundational concepts. Fear and anxiety are the twin pillars of aversive emotion.

Often, word origins help to clarify meanings. Not so with fear and anxiety. The word *fear* comes from the Old English *færan*, meaning to terrify with a sudden calamity. But the word also traces to Old High German *faren*, meaning to plot against. So fear seems to entail both immediate and impending danger. Things are no better with *anxiety*, which is rooted in the Latin *anxius* for a troubled mind, hence a sense of foreboding. However, the stem *anx* comes from the Latin *angere*, meaning to choke—a rather urgent circumstance.[3] Despite these linguistic ambiguities, psychologists have generally settled on common-sense descriptions of the emotions of fear and anxiety.

Fear is the heart-pounding response to present danger, and anxiety is the disquiet that comes with anticipating danger.[4] In clinical terms, the patient experiencing fear is highly aroused and seeks to escape the situation. In contrast, the anxious patient is worried and focuses attention on possible sources of impending harm. As such, fearful individuals exhibit physiological responses associated with taking flight, whereas anxious patients engage in preoccupied restlessness and negative self-talk.[5] We are anxious when descending the steps into the dark basement, and we are fearful when the light comes on to reveal a spider at our feet. So we can say that fear originates in a heightened physiological condition that elicits a mental state, and anxiety originates in the mind and gives rise to a bodily state.

Even when we know what object triggers a fear, it may be less clear exactly what danger is being perceived. Aaron Beck, the director of the University of Pennsylvania's Psychopathology Research Center that bears his name, distinguishes between proximate and ultimate fears, the former being the immediate stimulus and the latter being the consequence.[6] For example, a person who is frightened by cockroaches (proximate) might believe they will spread disease or invade her body (ultimate). Or a person who blanches amid a swarm of grasshoppers (proximal) might harbor an existential dread of being reduced to nothing or of being physically overwhelmed (ultimate). Moreover, a simple fear can "spread" into a pool of anxieties by association. The person who is afraid of cockroaches might become apprehensive about going into her basement or opening the cabinet under the sink. Likewise, a fellow frightened by a grasshopper swarm might harbor misgivings about entering gullies or encountering flocks of birds. And these anxious folks are not at all anomalous.

The prevalence of anxiety has risen since the 1960s to the point that this is now the single largest mental health problem in the United States. The most common ailments leading us to see a doctor are hypertension, cuts and bruises, and sore throat, followed closely by anxiety—which is well ahead of the common cold. Drugs to treat anxiety are among the most widely used medicines in the world, with billions of dollars being spent each year. Even with these interventions, only 10 percent of people with clinical anxiety are relieved of symptoms within four years of treatment (versus 80 percent of those with depression, who recover within two years).[7]

The socioeconomic cost of anxiety disorders (including extreme fears) is staggering, accounting for nearly one-third of the mental health care costs in the United States. In men, being highly fearful increases the risk of fatal coronary disease threefold and the risk of sudden death sixfold. Suicide rates in people with anxiety are equal to those in people suffering from depression, with the risk of suicide before the age of forty-five being five times higher than in the general population. Moreover, people with anxiety disorders are likely to self-medicate. Studies have found that one-half to two-thirds of alcoholics suffer from pathologically severe fears.[8]

* * *

Fear engages both our mental and our physical capacities. Imagine going into your laundry room.[9] Suddenly, you freeze. A huge spider is next to the washing machine. Your heart races and your body tenses (except your sphincters, which tend to relax). Then you realize that you've mistaken a wad of lint for a spider. You calm down and put the clothes in the washer.

What's interesting is that your perception of the "spider" *followed* your startle response—you didn't first think "There's a spider" and then prepare to bolt. Your body was ahead of your consciousness because the spider-like image was initially processed in the brain's thalamus which immediately passed on a crude representation of a leggy blob to the amygdala. Then the amygdala told your body to tense and to release epinephrine, insulin, and cortisol—a cocktail of hormones which increases your pulse, blood pressure, and respiration. Speed counts more than accuracy in responding to danger.

Fortunately, the thalamus copied the "Spidery thing!" message to your cortex for more thoughtful deliberation. The cortex made an accurate assessment, and you realized that you mistook a lint ball for a spider. This outcome was passed on to the amygdala and the alarm was canceled, allowing your skeletal muscles to relax and your sphincters to tense—thereby avoiding an unfortunate addition to the laundry basket. Hence, the response to scary things is a kind of neurophysiological "Ready, fire, aim" strategy.[10]

Of course, you might have reacted by stomping on the linty spider, rather than by bolting. The cognitive and physiological elements of fear would be similar, but your behavior would be "fight" rather than "flight." And there's a third option to this classic choice: freezing, which can manifest as fainting. This requires a different neurophysiological response—a sudden drop, rather than rise, in blood pressure (which means less bleeding if you're injured). Depending on the situation, survival chances might be best with any one of these responses. For example, wildlife experts tell us to pummel cougars (fight) and play dead with bears (freeze), while leaving the room (flight) seems appropriate for spider encounters. Recent studies have added a fourth option: "tend and befriend" (or "foster" to retain alliteration). Women are particularly inclined to look after one another when endangered.[11]

While we respond to fear in various ways, the feeling of fear—or at least what people report as their experience—is pretty much the same around the world. This is to be expected given the physiological foundation of this emotion. The symptoms of anxiety, however, differ markedly across cultures, as one might expect given that this emotion originates in our minds, which are far more susceptible to social forces. For example, 54 percent of anxious Indians have trouble sleeping, compared to 21 percent of Nigerians, and 45 percent of anxious Nigerians report accelerated heart rates, compared to 13 percent of Israelis. And although women are much more commonly afflicted with anxiety in Western cultures, in Eastern societies men seek out treatment for anxiety as often as do women.[12]

RAISING THE STAKES: PHOBIA

Paracelsus was a bombastic sixteenth-century polymath, celebrated today as the "father of toxicology." He is best known for recognizing that "all things are poison, and nothing is without poison; only the dose permits something not to be poisonous"—or, in modern parlance, "The dose makes the poison."[13] And this principle surely holds for the mind as well as for the body. In appropriate doses, anxiety improves performance and fear protects us. But in excess, anxiety becomes debilitating and fear transforms into phobia.

The word *phobia* originates with the god Phobos (whose name means "flight"—as in escape), and the Greeks put images of this deity on their shields to terrify the enemy. The first use of the term in a medical context was by the famed first-century scholar Celsus (Paracelsus—not lacking in self-confidence—gave himself this name because he thought he was "greater than Celsus"), who referred to hydrophobia (fear of water) as a symptom of rabies. However, psychologists did not start using the term *phobia* until the early 1800s.[14]

Today a phobia is understood as a "marked and persistent fear that is excessive or unreasonable, cued by the presence or anticipation of a specific object or situation." The gold standard for psychological evaluation is the *Diagnostic and Statistical Manual of Mental Disorders*, which lists five criteria for a person to qualify as being phobic.[15]

The first four benchmarks are rather noncontroversial. The experience of fear should: be reliable, persist for at least six months, induce significant impairment, and cause either avoidance of the stimulus or endurance with extreme distress. Put these together and you have a person whose life can be pretty awful. Consider the case of this poor fellow:

> One patient, for example, who had a great dread of being crawled over by cockroaches and other small insects, was always calm in public speaking, relished meeting new people, and was fearless in various athletic events. He felt extremely anxious whenever he was alone in his apartment at night because of his fear of being attacked by insects.[16]

And things didn't go so well for these unfortunate arachnophobes:

> An individual with a spider phobia who opened her car door and left the car after seeing a spider on her dashboard—while the car was moving. . . . A patient with a spider phobia who moved out of her home for several weeks after seeing several spiders over a period of a few days.[17]

Or for this entomophobe:

> On at least two occasions [the seven-year-old boy] had run across busy streets when confronted by a bee and his parents were worried that he might come to harm because of intense and sudden escape reactions.[18]

As odd as this might seem, grasshoppers had the same effect on Salvador Dalí, the brilliant and disturbed surrealist painter.[19] Along with melting watches, insects play an important role in Dalí's art. His paintings reveal fears and fantasies rooted in his youth—and insects are featured. The young and peculiar Dalí was mercilessly tormented by other children who threw grasshoppers at him. He became so hysterical that his teachers forbade their students to even mention grasshoppers. For the adult Dalí (whose abject terror of grasshoppers makes my response to swarms appear utterly normal), these insects became oversized, dreadful symbols of waste and destruction; he often depicted them as eating the main subject of a work. Also as a child Dalí discovered his pet bat dead and covered in ants (that he had a pet bat might explain something in itself), and so swarms of ants became symbols of mortality and decay. Dozens of his paintings include insects, and in *The Great Masturbator* Dalí incorporates both grasshoppers and ants into a single, profoundly disturbing painting.

At the age of thirty-seven, Dalí asserted, "Even today, if I were on the edge of a precipice and a large grasshopper sprang upon me and fastened itself to my face, I should prefer to fling myself over the edge rather than endure this frightful 'thing.'"[20] This was no vain threat, for he once jumped out a window in terror to escape from his insectan nemesis.[21]

Figure 1.2
The famous surrealist painter Salvador Dalí was emotionally traumatized by insects during his childhood, and these experiences provided the raw material for his works' deeply troubled expressions of his inner life; note the grasshopper in the center-left of the painting (image of Dalí provided by Library of Congress, Van Vechten Collection; image of Dalí's *Portrait of Paul Eluard* by Cea through Creative Commons).

A phobia can be emotionally, socially, and professionally devastating. Beck offers the example of an elevator phobia, which could "change the entire course of a person's vocational and personal life."[22] Even a subclinical phobia of being engulfed by insects could make an entomologist second-guess his career choice. The criterion of impairment is particularly relevant to a fear of insects, given that most entomophobic urbanites can largely avoid their psychological triggers—at least until the recent bed bug outbreak crystallized latent phobias and evoked dramatically irrational responses to these insects.

The final criterion of a phobia is that the individual recognizes the fear as unreasonable or excessive. However, recent studies have found that many people who meet the other criteria don't find their fears to be unwarranted. Interestingly, even when they do come to recognize that their fears are unfounded, their responses are often unchanged. According to Beck, "The anxious person enters a subjective world which is not readily understandable to an outside observer. This person's fears are, to him, totally reasonable."[23] This challenge to the diagnostic standard is supported by studies of arachnophobia.

People with an irrational fear of spiders "have relatively limited insight into the irrationality of their fears."[24] That is, being terrified strikes them as quite a reasonable response to these creatures. Several studies have found that, compared to those with a presumably sensible aversion to spiders, phobic subjects consistently offered higher estimates of the chances of being bitten by a spider, thought that a bite would cause catastrophic injury, and—here's the key—believed that their elevated levels of fear were highly appropriate (nonphobic individuals entertained the possibility that they had underestimated the risk).

In my favorite study, Australian scientists assessed phobic responses using "a tall glass cylindrical container with its open end uncovered, containing two dead Huntsman spiders."[25] They excluded subjects who figured out that the spiders were dead. Only two people evidently got close enough to discern the state of the spiders, which isn't surprising given that the creatures were six inches across. Phobic individuals and subjects in a control group were asked to estimate the probability of being bitten, the probability of being injured, and the reasonableness of their beliefs before, during, and after "exposure" (i.e., having the bejeebers scared out of them). The results showed that arachnophobes are pretty sure something bad will happen before and after an encounter; those in the study estimated a two- to fourfold greater probability of being bitten or injured than did the nonphobic controls. And they are really sure of harm while in the presence of a humongous spider; the arachnophobes' estimates of their likelihood of being bitten were eight times greater than those of the control subjects. What's more, they believe that their markedly greater fear is reasonable not only during the arachnid meet-and-greet but after the danger has passed. If phobics are supposed to be able to reflect on the excessiveness of their fear when not confronted with the stimulus, then somebody needs to tell the arachnophobes—or change the criterion.

Upping the ante from fear, we encounter panic.[26] In the parlance of poker, we are emotionally "all in." Panic is a sudden sense of terror that comes with a trembling, sweating, dizzying, hyperventilating, heart-pounding, overwhelming urge to escape. So intense are these feelings that people who have experienced panic often report a sense of dying, losing control, or going crazy, which makes the occurrence that much more awful—and likely exacerbates the physiological symptoms. Although panic attacks can occur unexpectedly, they can also be induced in phobic individuals through encounters with the feared objects. Those who have endured such events generally describe them as severe experiences of unmitigated terror.

* * *

Imagine that you're in a scavenger hunt organized by your therapist and "a phobic individual" is on your list—where would you look? If you just randomly ask people on a US street, you have about a one-in-ten chance of finding somebody with a phobia.[27] You would have a greater challenge in most other countries, although Canada and Iceland are better-than-average bets. If you want to up your chances, then look for a thirty-something, poorly educated, nonwhite, religious female.

Children often develop phobias between the ages of two and seven years, but they tend to outgrow them by early adolescence. However, late adolescents are potentially good choices. For animal phobias, the initial fear commonly begins around ten and then intensifies at about twenty years of age.[28] Across all phobias, the average age of onset is sixteen years, with excessive fears peaking between the ages of twenty-five and fifty-four. Once a person develops a phobia, it tends to hang around. Without treatment, the mean duration is more than twenty-two years. So someone in his or her early thirties should be just about right for your scavenger hunt.[29]

With regard to education, phobias are most common in adults with less than twelve years of schooling. Be aware in your search that there is an interaction with race; among whites, the highest rates of phobias are seen in those with zero to seven years of schooling, while prevalence in nonwhites is highest with eight to eleven years.[30] However, the overall pattern of more education, less fear, is generally borne out across studies.[31] If you can only find college-educated people on your scavenger hunt, then look for those who took biology courses. It turns out that about half of college freshmen who have taken biology are afraid of spiders, compared to only a third of those who haven't.[32]

African Americans have about twice the frequency of phobias as whites (19.7 percent versus 10.3 percent), and Mexican Americans are more likely to have phobias than non-Hispanic whites. During your hunt, remember that culture matters—phobias are 60 percent more common in Mexican Americans born in the United States than in those born in Mexico.[33] And pay attention to people's accents; phobias of animals, darkness, and bad weather are more than twice as prevalent in India as in the United Kingdom, although Brits have markedly higher rates of social phobias.[34] To further narrow your search for a phobic individual, be sure to look for religious jewelry during your hunt, as young adults who report that religion is very important to them are more likely to be phobic.[35]

And finally, don't waste your time asking men—the rates of phobia are twice as high in females (14.4 percent versus 7.8 percent). Besides, men lie. Or at least the studs do. Researchers have found that a high level of self-reported masculinity is a good predictor of a low level of self-reported fear. But when guys are told that they will be physiologically monitored for truthfulness, they report significantly greater levels of fearfulness.[36] Enculturation of traditional gender roles probably accounts for much of the difference. Stories, movies, and other images reinforce societal expectations that men should exhibit courage and women are expected to evince fear.[37]

Of course, the various factors associated with fear are not independent. For example, race and education are correlated, as are education and religiosity,[38] so it is difficult to tease apart which are the best predictors of phobias. In addition, we should not conflate correlations with causes. Being young, uneducated, religious, black, or female does not "make" one phobic, just as getting a tattoo won't cause you to prefer hot weather and developing a taste for mayonnaise won't turn you into a good dancer, despite the existence of correlations between these factors.[39]

THE EMOTIONAL WILD CARD: ENTOMOPHOBIA

What we fear has changed throughout history. In the sixteenth century, people were afflicted with the fear of demons manifesting as "satanphobia," and not until the nuclear age did we develop irrational fears of radiation.[40] Today, the most common phobias involve social settings (e.g., agoraphobia, the fear of being in places where escape is difficult or help is unavailable). Depending on the study, next comes one of the "specific phobias," phobias that are triggered by particular objects or situations (an earlier term was *simple phobias*). The most prevalent of these are induced by insects and other animals, tight spaces, and heights, followed by darkness, blood/injection/injury, flying, illness, and thunder and lightning.[41]

According to an extensive study, the otherwise awful 1970s song with the refrain "I don't like spiders and snakes" nailed the human condition.[42] These are the two most commonly feared animals, with about 6 percent of us having a debilitating fear of snakes and 4 percent suffering arachnophobia;

only acrophobia (fear of heights) and claustrophobia (fear of closed spaces) are more prevalent.[43] Psychologists have cataloged well over one hundred specific phobias.[44] Of these, thirty-eight involve animals (such as amphibians, birds, chickens, dogs, fish, and even otters). Insects and their relatives are featured in the following phobias:

Acarophobia: fear of insects that cause itching
Apiphobia or melissophobia: fear of bees
Arachnephobia or arachnophobia: fear of spiders
Entomophobia or insectophobia: fear of insects
Isopterophobia: fear of termites
Katsaridaphobia: fear of cockroaches
Mottephobia: fear of moths
Myrmecophobia: fear of ants
Pediculophobia or phthiriophobia: fear of lice
Scabiophobia: fear of scabies mites
Spheksophobia: fear of wasps

Entomologists can be nitpicky (literally and figuratively), and we grumble when people confuse insects with spiders, mites, or centipedes. However, it makes sense to use *entomophobia* as a kind of psychological wild card encompassing all terrestrial arthropods (lobsters, crabs, and shrimp are on their own). Studies of how the general public responds to insects and spiders reveal considerable confusion among respondents as to whether they are scared of things with six, eight, or lots of legs. Although some of us evidently differentiate among these creatures, many people simply feared small, leggy animals. Even psychologists studying phobias have adopted the vernacular of *bugs*, which presumably includes noninsects.[45] So if we take *entomophobia* to include insects and their common-sense relatives that people tend to lump together, then the United States is a country with nearly nineteen million entomophobes.[46]

When people were asked to score various arthropods in terms of anxiety-generating features (speediness, nearness, ugliness, sliminess, and suddenness of movement), the winners were spiders, followed by grasshoppers (it's not just me, evidently), ants, beetles, moths, butterflies, and caterpillars.[47] On the other hand, when individuals were asked to rate their affinity for various creatures, the least-loved arthropods were scorpions, followed by ants, crickets, spiders, bees, ladybird beetles, and butterflies.[48] The overall pattern seems reasonably consistent among studies, although how the question is posed and the choices of creatures affect the responses.

The various ways of perceiving insects suggest that both proximate and ultimate fears play important roles in entomophobia. Rather than being afraid of earwigs per se (proximate), a person might be afraid of bodily penetration (ultimate) thanks to the old wives' tale that these insects crawl into the ears of sleeping people and burrow into their brains. Entomophobia also might be confounded with a social fear of humiliation if one's weakness is exposed. Following my encounter with the grasshoppers, I became anxious (not clinically so, but to a discomfiting degree) that I might experience such an irrational response in the presence of students or peers.

* * *

In trying to understand and treat the fear of insects, psychologists have not been satisfied with lumping together all entomophobes. After all, fear comes in degrees, so scientists have come up with some nifty methods to figure out just how scared a person is. The standard approach is to measure how close an individual is willing to come to the feared object. A typical protocol works along these lines: in a windowless chamber the size of a small bedroom, a table is placed on which sits a clear container about the size of shoebox; the container holds a spider whose legs would stretch the width of a half dollar. Then it gets interesting:

> The patient was instructed to enter the room, walk up to the cage, remove the lid, insert one hand (or both), pick the spider up, and keep it in her hands for at least 20 sec. The importance of doing her very best was emphasized, but the patient was told, of course, that she was free to stop the test at any point.[49]

"Of course"—as if a full-blown arachnophobe would coddle a spider just because a guy in a white coat wanted to collect some data. Other researchers have used the approach of allowing the subject to remain stationary and telling her to pull a string to draw closer a clear box with a hefty spider inside. The distance that remains when the individual reaches her limit of proximity is a measure of fear.

Quantification of fear gets us closer to what's going on in entomophobia. However, more elegant techniques have allowed psychologists to push deeper into the infested mind—primarily the mind of the arachnophobe, who has become the "white rat" of research on human fear. And the results have revealed a remarkable inner world of faulty reasoning, distorted perceptions, and selective perspectives.

Uncovering the reasoning process of people with an extreme fear of spiders is a bit convoluted, but given the resulting insights as to the nature of phobia, it is worth our effort to delve into the research.[50] Scientists took thirty-one fearful and twenty-seven nonfearful women and presented them with syllogisms. These are the classical "puzzles" of logic. If all humans are mortal and Socrates is a human, what follows? Socrates must be mortal. The first clever move the researchers made in the study was to give the subjects two types of syllogisms: those involving nonscary animals and those involving spiders.

And then (I said this was complicated but pretty keen), they formulated these syllogisms in one of four ways: logically valid and emotionally believable (e.g., a spider is creepier than a fish; a fish is creepier than a pigeon; therefore a spider is creepier than a pigeon), valid and unbelievable (e.g., a pigeon is creepier than a fish; a fish is creepier than a spider; therefore a pigeon is creepier than a spider), invalid and believable (a pigeon is creeper than a fish; a fish is creeper than a spider; therefore a spider is creepier than a pigeon), and invalid and unbelievable (a spider is creepier than a fish; a fish is creepier than a pigeon; therefore a pigeon is creepier than a spider). The neutral syllogisms involved the sizes of flies, cats, and elephants.

Arachnophobes tended to fall for believable but invalid syllogisms, demonstrating a bias in their reasoning. They were more ready to accept a false conclusion if it aligned with their preconceptions. What's more, they were markedly slower to decide whether a syllogism was valid if there was a mismatch between validity and believability. You might expect that faulty reasoning would be most prevalent with the syllogisms involving spiders, but you would be wrong. The phobic women also struggled with neutral themes. The researchers speculated that belief bias may *precede* the development of arachnophobia, rendering these individuals less able to correct their erroneous beliefs about the harmfulness of spiders.

Not only do arachnophobes have problems reasoning, but they also labor under an aberrant perception of reality. Compared to nonphobic individuals in the presence of spiders, phobics describe the creatures as bigger, uglier, jumpier, and faster. In one experiment, fearful individuals, compared to controls, rated the spider as being three times as likely to come out of its glass bowl, twice as likely to make an unpredictable movement, and three times as likely to move toward them.[51]

These differences are manifest even with imaginary spiders. In a classic study, individuals were first given the MSLQ—the Modified Spider Looming Questionnaire. Psychologists have tests for nearly everything, and this one includes questions pertaining to an individual's preoccupations (e.g., worrying about spiders), avoidance-coping (e.g., getting others to remove spiders), and vigilance (e.g., checking for spiders before sitting down). The subjects were divided into high- and low-fear groups and asked to simply imagine themselves in a room shared with a spider and three other people. The fearful individuals were significantly more likely to imagine the spider as angry and belligerent—intentionally singling them out and moving rapidly toward them. The researchers concluded that the distorted world of highly fearful individuals produces a heightened sense of impending danger.[52]

In addition to perceiving the world differently, phobic individuals also use experience to selectively reinforce their fears. All people use different tactics to test their rules about safe and dangerous stimuli. When it comes to rules about safe objects, one adverse experience is often enough to overturn a

Figure 1.3
Even inside a container, spiders evoke dread in arachnophobes, who describe these creatures as being larger as well as uglier, more aggressive, and nimbler than do nonphobic individuals (image by Matt Reinbold through Creative Commons).

person's previous presumption. A single wasp sting might be all it takes for a child to abandon the notion that prettily colored things are harmless. Conversely, if we have a rule that a class of objects is hazardous, we demand considerable evidence to overturn our presumption. A practical joker with a rubber spider doesn't convince us that we needn't avoid spiders.

What differs between normal and arachnophobic people is that for the latter, even the misperception of a threat can serve as evidence to confirm their rule about danger.[53] So when a spider does no damage other than to elicit psychological distress, this counts in the mind of the fearful individual as a harmful event and thereby reinforces—rather than refutes—the belief that spiders are dangerous. But the tendency for what psychologists call "confirmation bias" runs even deeper.

Researchers told twenty arachnophobic subjects that they would view pictures of spiders, weapons, and flowers, and that in some cases an electrical shock would follow an image. Before the experiment was conducted, the subjects anticipated that there would be about a three-in-four chance of receiving a shock in connection with either a weapon or a spider. Then they were shown a series of images that were randomly paired with electrical shocks so that there was actually a one-in-three chance that any photograph would come with a shock. As the experiment unfolded, the subjects corrected their initial impression and reported that they were experiencing a shock 32 percent of the time in association with the image of a weapon; they had processed the disconfirming evidence and modified their view of the world. However, despite the same evidence, they were unable to fully correct their preconceived bias about spiders and pain, reporting that they were being shocked 57 percent of the time when a spider image was shown.[54]

A BAD BET: ENTOMOPHOBIA IN THE MODERN WORLD

In 1950, O. Hobart Mowrer, who was to become president of the American Psychological Association, struggled to make sense of phobias. He described them as "the paradox of behavior which is at one and the same time self-perpetuating and self-defeating!"[55] The entomophobic individual lives with one foot in the real world of rational concerns. After all, insects *can* invade our space, evade us, reproduce rapidly, and then feed on everything from our joists to our blood. Granted, most insects are not nefarious. But some are—and the entomophobe's other foot is planted in a nightmarish world of amplified biological reality. Through poor reasoning, myopic attention, selective memory, biased interpretation, resistant convictions, and misperceptions of size, speed, and intentions, the entomophobe digs the hole of fear deeper. But such an individual's world may not be that far from our own.

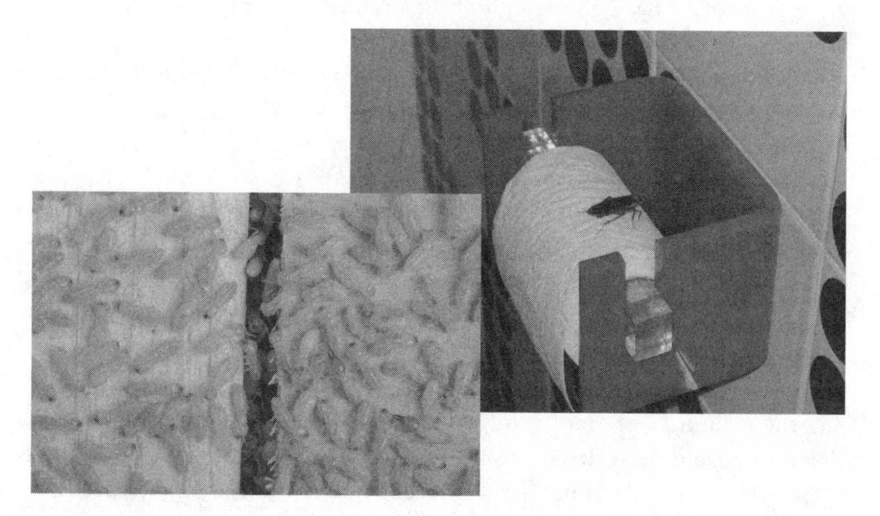

Figure 1.4
Insects evoke fear in humans through their capacity to invade, evade, reproduce, harm, disturb, and defy us—qualities that are evoked through even fleeting encounters with creatures such as cockroaches and termites inside our homes (images by Ted and Dani Percival through Creative Commons and US Department of Agriculture).

A survey of more than one thousand households revealed that 38 percent of the respondents disliked arthropods in their yards and 84 percent did not want them inside their homes.[56] So an aversion toward insects, if not full-blown entomophobia, is extremely common in our society. While most people would not refuse a job in the southern United States because of the region's bugginess (as the son-in-law of a good friend of mine contemplated, much to my friend's dismay), our antipathy toward insects has a cost. We should not dismiss the suffering of the 19 million entomophobic Americans, but our society surely pays a higher price for the other 265 million people who are just plain antsy.

True entomophobes may saturate their environments with poisons to protect themselves,[57] but even without phobic motivation, the first resort of many people to being confronted with insects is to spray insecticides. Such a tendency is particularly prevalent in urban situations, according to a classic 1976 study of how people respond to insects.[58] Along with the overuse of insecticides—a practice that is hardly rare in agricultural settings either—come health risks, contamination of water and soil, elimination of natural enemies, insecticide resistance, and resurgence of pest populations.

The urban study mentioned above included an account of a gardener who used a high-dose mixture of three pesticides to wipe out the whiteflies on his tomatoes. The witch's brew included two insecticides and a fungicide (perhaps the whiteflies looked a bit like mold, so he threw in the fungicide for good measure). The fellow had more than enough tomatoes for his needs, but he was revolted by the possibility that the insects might contaminate his food.

The belief that insects spread filth resonates with American fastidiousness. Antibacterial soaps, virus-killing tissues, and No-Pest Strips promise to keep our homes sterile. Recent outbreaks of food-borne illness have added pathogenic fuel to the decontamination fire. And, as noted in the 1976 paper, the government does not help by issuing food sanitation standards that refer to both rodent and insect "filth," treating rat feces on a par with insect fragments—an equivalence that persists in today's guidelines.[59]

Although insects can decimate crops and spread disease, the overwhelming majority are harmless or beneficial. But our irrational fears drive us to "protect" ourselves from innocuous species by poisoning our homes, polluting the environment, and throwing out perfectly good food. How we have come to harbor such dread toward insects is a fascinating and important riddle. If we can understand the origins of our fear, then we might effectively treat the millions of people with entomophobia—and reshape how the rest of us perceive insects. In so doing, we might substantially improve the quality of everyone's lives.

NOTES

1. Ed Diener, *Culture and Well-Being* (New York: Springer, 2009), 211.
2. Robert Plutchik, *Emotion: Theory, Research, and Experience*, vol. 1, *Theories of Emotion* (New York: Academic Press, 1980).
3. Aaron T. Beck, Gary Emery, and Ruth L. Greenberg, *Anxiety Disorders and Phobias: A Cognitive Perspective* (New York: Basic Books, 2005), chap. 1.
4. Ibid.
5. Ibid.
6. Ibid., chap. 6.
7. David H. Barlow, *Anxiety and Its Disorders: The Nature and Treatment of Anxiety and Panic*, 2nd ed. (New York: Guilford Press, 2002), chap. 1.
8. Ibid.
9. This story and the neurophysiological account are adapted from Ulrike Rimmele, "Emotions and learning," Caring Awareness, http://caringawareness.com/resources/library/emotional/E_Learn.html (accessed October 25, 2011).
10. The view that emotions are perceptions of certain bodily states was originally proposed by William James and Karl Lange and has been reinvigorated by Jesse J. Prinz in *Gut Reactions: A Perceptual Theory of Emotion* (New York: Oxford University Press, 2006).
11. Scott O. Lilienfeld, "Fear: Can't live with it, can't live without it," *Phi Kappa Phi Forum* 90 (Fall 2010): 16–18.
12. Ibid.
13. "Paracelsus," Wikipedia, http://en.wikipedia.org/wiki/Paracelsus (accessed October 25, 2011).
14. Beck et al., *Anxiety Disorders and Phobias*, chap. 1.
15. American Psychiatric Association, *Diagnostic and Statistical Manual of Mental Disorders*, 4th ed., text revision (Arlington, VA: American Psychiatric Association, 2000).
16. Beck et al., *Anxiety Disorders and Phobias*, 117.
17. Martin M. Antony and David H. Barlow, "Specific phobias," in Barlow, *Anxiety and Its Disorders*, 380.
18. S. Rachman, *Phobias: Their Nature and Control* (Springfield, IL: Thomas, 1968), quoted in Tad N. Hardy, "Entomophobia: The case for Miss Muffet," *Bulletin of the Entomological Society of America* 34 (1988): 65.
19. Meredith Etherington-Smith, *The Persistence of Memory: A Biography of Dalí*, (New York: Random House, 1992); Salvador Dalí, *The Secret Life of Salvador Dalí* (1942; repr., Whitefish, MT: Kessinger, 2010).
20. Dalí, *The Secret Life of Salvador Dalí*, p. 128.
21. Michael Elsohn Ross, *Salvador Dalí and the Surrealists: Their Lives and Ideas* (Chicago: Chicago Review Press, 2003).
22. Beck et al., *Anxiety Disorders and Phobias*, 122.
23. Ibid., 26.
24. Mairwen K. Jones and Ross G. Menzies, "Danger expectancies, self-efficacy and insight in spider phobia," *Behaviour Research and Therapy* 38 (2000): 585.
25. Ibid., 588.
26. Barlow, *Anxiety and Its Disorders*, chap. 4.
27. Ibid., chap. 1; Beck et al., *Anxiety Disorders and Phobias*, chap. 6.
28. Hardy, "Entomophobia."

29. Antony and Barlow, “Specific phobias”; Barlow, *Anxiety and Its Disorders*, chap. 1.
30. William W. Eaton, Amy Dryman, and Myrna M. Weissman, “Panic and phobia,” in *Psychiatric Disorders in America: The Epidemiologic Catchment Area Study*, ed. Lee N. Robins and Darrel A. Regier (New York: Free Press, 1991).
31. David N. Byrne, Edwin H. Carpenter, Ellen M. Thoms, and Susanne T. Cotty, “Public attitudes toward urban arthropods,” *Bulletin of the Entomological Society of America* 30 (1984): 40–44.
32. Hardy, “Entomophobia.”
33. Antony and Barlow, “Specific phobias.”
34. Ibid.
35. Ibid.
36. Ibid.
37. Of the first one hundred images of people generated by a Google search using the keyword “courage,” seventy were of men, while an image search for “afraid” yielded sixty-seven pictures of women in the first hundred. The role of the media in shaping our fears will be explored more deeply in chapter 3.
38. “Education,” National Association for the Advancement of Colored People, http://www.naacp.org/programs/entry/education-programs? (accessed November 13, 2011); Bruce Sacerdote and Edward L. Glaeser, *Education and Religion*. Harvard Institute of Economic Research Discussion Paper No. 1913 (Cambridge, MA: National Bureau of Economic Research, 2001).
39. Gus Lubin and Tony Manfred, “10 crazy correlations between unrelated things,” Business Insider, June 17, 2011, http://www.businessinsider.com/crazy-correlations-2011-6 (accessed November 13, 2011).
40. Beck et al., *Anxiety Disorders and Phobias*, chap. 7.
41. Ibid.; Antony and Barlow, “Specific phobias.”
42. Jim Stafford, “Spiders and Snakes,” lyrics available at Oldie Lyrics, http://oldielyrics.com/lyrics/jim_stafford/spiders_and_snakes.html (accessed October 25, 2011).
43. Eaton et al., “Panic and phobia”; Antony and Barlow, “Specific phobias.”
44. Jack D. Maser, “List of phobias,” in *Anxiety and the Anxiety Disorders*, ed. A. Hussain Tuma and Jack D. Maser (Hillsdale, NJ: Lawrence Erlbaum, 1985); Fredd Culbertson, “The phobia list,” http://phobialist.com (accessed October 25, 2011).
45. Eaton et al., “Panic and phobia.”
46. Antony and Barlow, “Specific phobias.” This estimate is based on the conservative assumption that half as many people fear insects as fear spiders.
47. Jamie Bennett-Levy and Theresa Marteau, “Fear of animals: What is prepared?,” *British Journal of Psychology* 75 (1984): 37–42.
48. Byrne et al., “Public attitudes toward urban arthropods.”
49. Lars-Göran Öst, Paul M. Salkovskis, and Kerstin Hellström. “One-session therapist-directed exposure vs. self-exposure in the treatment of spider phobia,” *Behavior Therapy* 22 (1991): 411.
50. Peter J. de Jong, Anoek Weertman, Robert Horselenberg, and Marcel A. van den Hout, “Deductive reasoning and pathological anxiety: Evidence for a relatively strong ‘belief bias’ in phobic subjects,” *Cognitive Therapy and Research* 21 (1997): 647–62.
51. S. J. Rachman and M. Cuk, “Fearful distortions,” *Behaviour Research and Therapy* 30 (1992): 583–89.
52. John H. Riskind, Roger Moore, and Laurie Bowley, “The looming of spiders: The fearful perceptual distortion of movement and menace,” *Behaviour Research and Therapy* 33 (1995): 171–78.

53. Peter J. de Jong, Birgit Mayer, and Marcel van den Hout, "Conditional reasoning and phobic fear: Evidence for a fear-confirming reasoning pattern," *Behaviour Research and Therapy* 35 (1997): 507–16.
54. Peter J. de Jong, Harald Merckelbach, and Arnoud Arntz, "Covariation bias in phobic women: The relationship between a priori expectancy, on-line expectancy, autonomic responding, and a posteriori contingency judgment," *Journal of Abnormal Psychology*, 104 (1995): 55–62.
55. Quoted in Barlow, *Anxiety and Its Disorders*, 10.
56. Byrne et al., "Public attitudes toward urban arthropods."
57. Hardy, "Entomophobia."
58. Helga Olkowski and William Olkowski, "Entomophobia in the urban ecosystem: Some observations and suggestions," *Bulletin of the Entomological Society of America* 22 (1976): 313–17.
59. US Food and Drug Administration, "Defect levels handbook: The food defect action levels," http://www.fda.gov/Food/GuidanceRegulation/GuidanceDocuments RegulatoryInformation/SanitationTransportation/ucm056174.htm (accessed October 25, 2011).

CHAPTER 2

Evolutionary Psychology: Survival of the Scaredest

I like to believe that I have more in common with Charles Darwin than with a monkey. But Darwin's account of his playful experiments at the London Zoo gives me cause to wonder. He had read that monkeys would go ape in the presence of snakes. So, like a kid who has to touch a bench with a "Wet Paint" sign, he took a stuffed snake into the monkey house: "The excitement thus caused was one of the most curious spectacles which I ever beheld. [The monkeys] dashed about their cages, and uttered sharp signal cries of danger, which were understood by the other monkeys."[1] He also tried out a mouse, a turtle, and a dead fish, but the monkeys were unperturbed. Their reaction was snake-specific. Being the consummate wonderer, Darwin continued his scientific shenanigans:

> I then placed a live snake in a paper bag, with the mouth loosely closed, in one of the larger compartments. One of the monkeys immediately approached, cautiously opened the bag a little, peeped in, and instantly dashed away. [Then] monkey after monkey, with head raised high and turned on one side, could not resist taking a momentary peep into the upright bag, at the dreadful object lying quietly at the bottom.[2]

Darwin was sixty-two years old when he played this somewhat adolescent prank on his primate pals, so at least he and I share an abiding sense of child-like curiosity. It is the monkeys' response to the snake-in-the-bag experiment that most intrigues me. For you see, a week after my heart-pounding encounter with the grasshopper swarm, I returned to Whalen Canyon. Scientific protocol required that somebody revisit the nearby experimental plots, although heading back to the seething infestation wasn't required. But like the monkeys, I couldn't resist. Maybe I was trying to prove to myself that I was in control—or maybe the promise of repeating the dark thrill of a primal experience was irresistible. This time, however, rather than descending into the recesses, I walked along the crumbling edge of the draw, peeking into the dreadful carpet of grasshoppers lying at the bottom.

ANXIOUS GENES

We credit Charles Darwin with the discovery of biological evolution as evidenced by giraffes' necks and finches' beaks. However, the great scientist understood that not only anatomy but also behavior—emotions, thoughts, and perceptions—has been shaped by the forces of natural selection. The underlying concept of evolutionary psychology is rather straightforward. The ways in which we feel, think, and perceive have consequences for whether we survive and reproduce. And to the extent that our mental abilities and predispositions are heritable, our descendants will share our psychological proclivities.

When it comes to objects that are potentially dangerous, evolution favors anxious genes.[3] That is, our ancestors were better off to err on the side of caution. Mistaking a twisted stick for a snake, a tumbling leaf for a spider, or a grass seed for a louse would have been better than ignoring these cues. A "false positive" meant an unnecessary flinch or some pointless scratching, while a "false negative" meant elimination from the gene pool. Fine-tuning anxiety continues to be our challenge. With too little anxiety a boxer drops his guard, and with too much anxiety a dancer suffers stage fright. Indeed, moderately anxious animals perform better on simple tasks—such as dodging slithering snakes and surreptitious spiders—than do those that are not anxious.[4] From the perspective of evolutionary psychology: "The cost of survival of the lineage may be a lifetime of discomfort."[5]

Our evolutionary history as soft, slow sources of protein and vulnerable targets of venom quite reasonably accounts for our tendency to be alarmed by creatures that can eat, sting, or bite us. Cultural and technological changes happen much faster than genetic change, so we are left with minds and bodies poised for dangers on the savanna while we try to stay safe on the freeway. Our ancestors roamed a world of lions, spiders, and snakes for about four million years, while we've wandered the gritty streets of civilization for no more than ten thousand years. If we think of the history of our genus (*Homo*) condensed into a single year, we started building cities at 2 A.M. on December 31.

So it will be a while before we evolve the psychological tendency to fear guns and automobiles in proportion to their likelihood of killing us. In the United States, this deadly duo of modern technology is involved in about six hundred times more deaths than all animal-related fatalities.[6] Even so, our contemporary fears of various animals are not entirely misguided. Entomophobes are onto something given that the animals most likely to kill you are the stinging insects (bees, wasps, and hornets are responsible for about 50 deaths per year in the United States)—primarily through allergic reactions. Aside from work-related deaths involving animals (cattle and horses being dangerous on account of their size), the next most lethal animals are dogs (19 deaths per year) followed by spiders and snakes (both accounting for around 6 deaths per year).[7] These statistics don't include the combination of machines and animals; vehicle collisions with deer, livestock, and other creatures account for 165 US deaths per year—more than all the other animal-related fatalities combined.[8]

As for predators, save your adrenaline. In the past century, bears, mountain lions, and sharks have killed only 77 people in the United States (on par with the number killed by firearms or vehicles in a typical day).[9] However, there is one ancient fear that still makes sense—much of humanity has reason to be worried about bloodthirsty insects. Malaria causes nearly ten times more deaths around the globe than all of the vehicles and guns in the United States.[10] So it is that we are the descendants of hominids with a psychological disposition to dodge predators—and to swat mosquitoes.

BORN SCARED: HARD-LINER EVOLUTIONARY PSYCHOLOGY

The strong view of evolutionary psychology is that our fears are innate—we do not need to learn avoidance of spiders and roaches and bears (to adapt Dorothy's short list of dangers in the forest of Oz). After all, learning would be for losers if it meant that early humans had to survive a tiger attack or cobra bite in order to associate the animal with danger. Nor would prehistoric children live long enough to complete homeschooling. Our survival depended on our obeying genetically ingrained, preverbal rules. We reacted decisively rather than thinking, "Skittering things are dangerous." These inborn commands were overly inclusive for a good reason: avoiding a wad of spider-like hair on the cave floor had nominal costs, while grasping a vine-like snake in the jungle was disastrous.

Our contemporary fear of spiders might be rooted in the African savanna, thanks to species such as the six-eyed sand spider—a creature that buries itself in a thin layer of sand, detects the vibrations of passing prey, and leaps out to deliver a dose of potent venom (the prey is usually an insect, but mistakes are possible when the spider can't see what's on the menu until the moment of attack).[11] Likewise, experts contend that an innate fear of insects is the legacy of our ancestors, who ran a gauntlet of six-legged assailants that spread diseases through both contact (a typical house fly in a modern slum is coated with four million bacteria)[12] and blood feeding (from bubonic plague to yellow fever), along with delivering toxic and allergenic chemicals through stings and bites.

Springing spiders and filthy flies make for good speculation, but what is the evidence for genetically encoded fear? Scientists have found a heritable disposition of emotionality that underlies anxiety and depression, and some contend that we inherit a nonspecific tendency to be fearful.[13] Other researchers argue for a stronger position: that we inherit particular classes of fear. This view is supported by studies showing that the heritability of animal fears, for example, is 47 percent (the highest such heritability is seen with agoraphobia at 67 percent).[14] And twin studies, which allow researchers to isolate genetic and environmental factors, have revealed that specific phobias are perhaps the only psychological disorders for which there is compelling evidence of a direct genetic contribution.[15] Animal fears, along with so-called BII (blood, injection, and injury) fears, are particularly prone to run in families. Children of parents with such fears run three times the risk of having the same specific phobias.[16] However, nobody is suggesting that there is a single gene for entomophobia, as there is for albinism or dwarfism.

Experimental data also support the possibility that we are born scared. Although it is difficult to separate instinct from learning, people consistently overestimate the likelihood that fear-relevant stimuli (e.g., an image of a spider) will be associated with a painful experience (e.g., an electric shock) compared to neutral stimuli.[17] More compellingly, a fearful response may not even require awareness of the aversive creature. When subjects were presented with various images for just thirty milliseconds (faster than the blink of an eye) followed by a "masking stimulus" (to assure there would be no conscious recognition), arachnophobes showed a marked change in skin conductance indicating fear.[18] Further studies along these lines provide evidence that our responses to dangerous animals arise from a "specifically evolved primitive neural circuit that emerged with the first mammals long before the evolution of the neocortex."[19] In other words, we are hardwired for fear.

However, we are not genetically imprinted with a detailed bestiary. The sinuous shape of a snake—rather than the details of any particular serpent—is crucial to our rapid detection of these legless animals.[20] As for six- and eight-legged dangers, young mammals, including humans, keenly attend to characteristics such as skittering movements.[21] Indeed, it has been proposed that our reaction to insects results from their erratic motion, which leads to retinal images similar to those involved in falling—and this causes a startle response that we then interpret as fright.[22]

Psychologists have found that fear of motionless insects and spiders is also triggered by visual cues (the "ugliness" of antennae projecting from the head and the oddly proportioned eyes and bodies) to which we add tactile features (the discomfort of hairiness and sliminess).[23] It seems that the more divergent a creature is from the human form, the greater its capacity to evoke fear. According to Eric Brown, a scholar of science and literature, "[It] is partly this innate distortion, derangement, and fragmentation—their world ever

Figure 2.1
People find insects frightening, particularly when they have odd projections, bizarre eyes, hairy bodies, and strange proportions. Even rather dumpy creatures such as this beetle can elicit fear with dangling, paddle-like antennae, beady eyes, hunched back, furry thorax, and diminutive head (image by Malham Tarn through Creative Commons).

fractured from our own—that marks insects as humanity's Other."[24] And this sense of alien life, with all of its symbolic and imaginative implications, has spawned strange and enticing psychological theories regarding the primal roots of our fear.

DEEP MEMORIES OF INSECTS

According to Maurice Maeterlinck, the 1911 Nobel laureate in literature, "No matter what monsters have defiled or terrified the surface of the globe, we bear them within us. . . . they are only awaiting an opportunity to escape from us, to reappear, to reconstitute themselves, to develop, and to plunge us once again into terror."[25] The idea that monsters lurk deep within the human psyche draws upon the psychological work of Sigmund Freud and Carl Jung. Subsequent scholars have used the Freudian-Jungian framework to make sense of our response to insects: "The cockroach is an archetypal image. . . . Western cultures relegate it to the darkness [and] associate [it] with the unconscious and the power of the id."[26] Just as Darwin didn't understand the mechanism of inheritance but understood its importance to the origin of species, neither Freud nor Jung grasped how insects had infested the human psyche. Both understood, however, that our thoughts about these creatures were deeply embedded, even if the origin of these thoughts awaited the development of evolutionary psychology.

Sometimes a cigar is just a cigar, but a cigarette beetle (*Lasioderma serricorne*) might have been another matter to Freud. For the great psychologist, insects served as important symbols; his patients drew connections between these creatures and their own psychological problems. And fear played a central role. For example, Freud proposed that the fear of being bitten by a spider represented the fear of being punished by one's father (today's entomologically enlightened psychoanalysts favor the spider appearing in dreams as symbolizing a devouring or castrating mother, given that some female spiders cannibalize their mates).[27]

Believing that the mental associations of insects with neuroses developed during the life of an individual, Freud delved into the histories of his patients. In a classic case of a disturbed young man, the poor fellow recounted that the "opening and shutting of the butterfly's wings while it was settled on the flower [looked] like a woman opening her legs."[28] With evident delight, Freud hastened to note that the sticklike projections from the swallowtail's wings "might have had the meaning of genital symbols."[29] Sure enough, with some further encouragement the fellow recalled a time in his youth when his nursemaid—whom the boy had ogled while she was inadvertently positioned with her legs apart—told him that masturbation would cause his penis to fall off, which he figured would render him like his sister, whose genitals he'd glimpsed. The case history gets a lot weirder, but we'll leave it here.

Freud was intrigued and perplexed by the frequency with which insects were the subject of his patients' phobias.[30] Either there was a remarkable coincidence between these particular creatures and traumatic events in people's lives or some much deeper sort of memory—not dependent on individual experience—was being accessed. And the latter explanation became the life's work of Freud's successor.

Carl Jung can be viewed as the forerunner of evolutionary psychology. Although such an interpretation of intellectual history is unconventional, the connection has been made by contemporary researchers: "It is not unreasonable to assume that the danger and annoyance that insects have caused to man over the millennia has resulted in an ingrained fear of insects. . . . An almost Jungian fear of insects can therefore be rationalized in all of us."[31] Jung called these primal, collective memories "archetypes"—the "inherited memories from the evolutionary past of the race." Not only did Jung use insects as examples of this innate knowledge, but it has also been said that for him, "the unconscious was an insect."[32]

Jung proposed that our basic drives (Freud's so-called id) could be seen in the evolved behavior of insects. Perhaps he was persuaded by Freud's having posited that the phenomenon of transference in humans (the unconscious redirection of feelings from one person to another) was a vestige of the psychic—even telepathic—communication employed by ants, which we now attribute to pheromones.[33] So it was that Jung argued for humans to heed instinctual knowledge, noting that "if man sometimes acted as certain insects do he would possess a higher intelligence than at present."[34]

The most famous insect-derived insights of Jung's career came during a counseling session he had with a woman who was no longer making progress in therapy. As he listened to her describe a dream about having been given a golden scarab—the ancient Egyptians' sacred symbol of rebirth—an insect flew in through an open window. It was a local species of scarab beetle, and its arrival so amazed his patient that she was able to break the impasse and resume her therapeutic growth.[35]

Such connections between the dream and waking worlds were vital to Jung's work. He maintained that we can come to know the deep wisdom of the unconscious through archetypal symbols that we encounter in our dreams. For example, he asserted that a spider represented dread of the unconscious forces of integration, and a queen ant was the female equivalent of a father who maintains his daughters as puppets by impeding their sexual and emotional maturation—an entomologically informed, if somewhat bizarre, interpretation.[36]

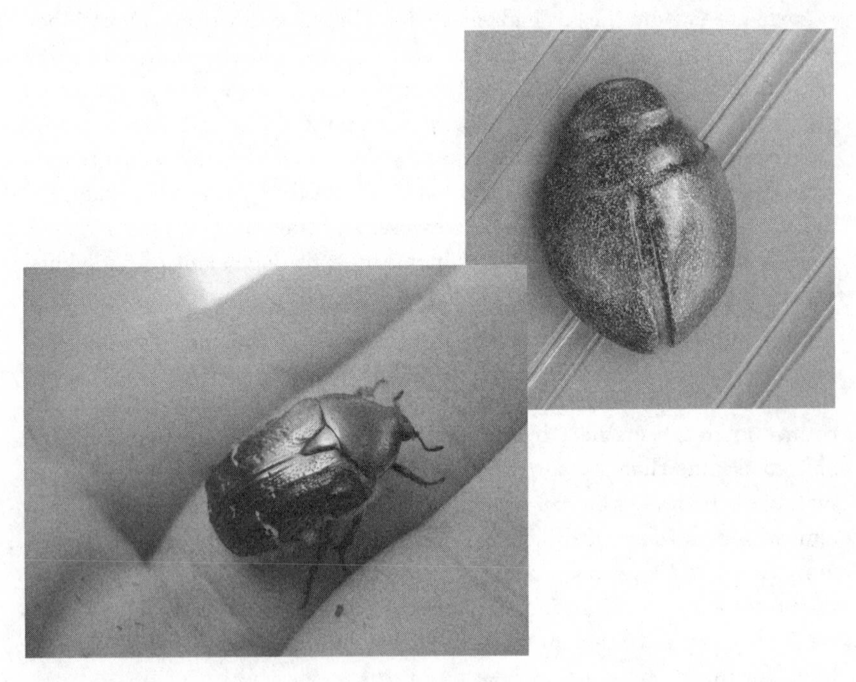

Figure 2.2
Carl Jung's most famous entomological therapeutic breakthrough with a patient involved her dream of a scarab beetle (which served as a symbol of rebirth for the Egyptians, who carved its image into amulets) and the serendipitous arrival of a local member of this family of insects, which are common garden pests (images by Ficusdesk through Creative Commons and Bugman95 through Wikimedia Commons).

The Jungian archetype spawned a cottage industry of dream interpretation that persists today—and insects play featured roles. Dream dictionaries invariably have several entries for insects.[37] Only a few of these are positive. Butterflies and crickets fare well as symbols of the soul and happiness. Caterpillars mean unrealized potential—which could be a good thing or a bad thing. More complicated manuals contextualize the insects, so that an ant signifies being "bugged" by something at an emotional level but represents collective harmony at a spiritual level. Likewise, spiders and bees can go either way.

Most insects in dreams, according to modern interpreters, are bad news. The unconscious mind is troubled when cockroaches, flies, and maggots show up. A passage from a lyric essay by a poetically inclined anthropologist evokes my waking-dream experience within that grasshopper-filled draw:

> There is the nightmare of fecundity and the nightmare of multitude. There is the nightmare of uncontrolled bodies and the nightmare of inside our bodies and all over our bodies. . . . There is the nightmare of swarming and the nightmare of crawling. . . . There is the nightmare of beings without reason. . . . There is the nightmare of too many limbs.[38]

The mantle of Jungian authority was passed to James Hillman, who carried the torch for archetypal psychology until his death in 2011. Even more ethereal than Jung in his views, Hillman maintained that the human soul is a repository of ancient wisdom: "Our dreams recover what the world forgets."[39] He collected his patients' dreams and classified them along four themes. Setting aside concerns about confirmation bias, sampling problems, and other methodological shortcomings, Hillman's accounts are enchanting. First, he saw insects as representing hidden intentions, as with the woman who dreamed of a colorful, human-sized hornet that grasped her finger and pulled her away from her home, but despite her fascination with him (the hornet was evidently male), she decided to stay to care for her child. Next there were wound dreams, with insects emerging as warnings, the precise nature of which depended on the species. Third, eradication dreams entailed a multiplicity of ways to destroy insects (e.g., fire, crushing, scissors)—and the psyche's destructive impulse was manifest in our waking hours as the egregious misuse of chemical insecticides. Finally, the element of mystery was manifest in dreams such as that of a huge mantis asking a woman if she was a citizen, causing her to awaken and scream, "No!" Hillman took this to be a cosmic challenge to his patient regarding her separation from, and duties to, the natural world.[40]

Not many evolutionary psychologists would count psychoanalysts among their ranks, but the two groups share an understanding that we possess inborn proclivities. The latter group might have missed the mechanism accounting for these tendencies, but the history of science is often a matter of being right for the wrong reasons.

THE LANGUAGE OF FEAR: EVOLUTIONARILY PREPARED TO LEARN

Critics of hard-line evolutionary psychology note that the objects of phobias often don't occur in nature (e.g., clocks, clowns, and computers), so these particular fears could not have been selected through evolution.[41] They are right, but even if some scientists overstate the case that our fear of insects is hardwired, it is clear that the human mind is not a blank slate. We are born with tendencies to readily learn things that favor our survival—such as language. Nobody emerges from the womb able to speak English or Mandarin, but as we mature our minds are strongly predisposed to pick up the grammar and vocabulary being used around us. We are evolutionarily prepared to learn languages—and fears. English and Mandarin fit our inborn expectations of language structure, while wasps and cockroaches fit our innate template of fearful objects.

Researchers have proposed the existence of various "fear modules" (similar to linguistic and moral modules)[42] that evolved to guide our thinking, learning, and behaving.[43] For example, we appear to have a propensity to discern predators (which allowed us to avoid being attacked by big animals), social hierarchies (which allowed us to avoid being throttled by dominant individuals), and contamination (which allowed us to avoid contracting disease from filthy things and sick people). We are attuned to general concepts that are manifest in particular situations or objects. We are predisposed to a fear of falling, and airplanes fit the bill. If one is predisposed to fear skittering objects, bodily invasion, and suffocating sensations, then a swarm of grasshoppers fits the bill.

Other scientists take a softer view, suggesting that we inherit merely a propensity for a particular level of fear, with too little making us dangerously bold and too much making us impotently hesitant. The concept goes back to the *Nicomachean Ethics*, in which Aristotle proposed that the virtues lie between the extremes, with courage being the "golden mean" between foolhardiness and cowardice. So the development of a phobia is a matter of extremes—being overly prepared to learn a kind of fear and experiencing stimuli as overly potent.[44]

According to this view, entomophobic individuals are genetically primed for defensive tendencies, such as a low perceptual threshold for being startled by peripheral movements or a hyperactive autonomic nervous system that dumps adrenaline at the drop of a hat—or the flutter of a wing.[45] These biological proclivities interact with environmental cues (e.g., encounters with moths fluttering at the edge of our visual field) to rapidly generate our learning of fearful responses to specific cues. And the learning need not be cognitive. That is, we don't necessarily encode our fear as declarative, verbal knowledge: "Fluttering moths scare me." Emotional learning does not depend on conscious awareness, so we may not be able to articulate or think clearly about why we flail at a flapping moth.[46]

While "prepared learning" sounds like a reasonable way of understanding our fear of insects, evolutionary accounts can be nothing more than "Just-So stories." A plausible tale of biological adaptation can be invented for almost any structure or behavior. Fortunately, there is empirical support for our having entomophobic predispositions.

Some intriguing evidence comes from our primate relatives. Like the monkeys in the London Zoo, many species exhibit fear of snakes. And compared to the relatively sedate responses of lab-reared primates, the intensity of alarm calls and mobbing behavior in the wild resembles full-blown panic. This difference is a matter of experience, and experiments have shown that lab-reared monkeys are exquisitely primed to acquire certain fears. Laboratory monkeys that had never seen a snake were rapidly conditioned to respond with fear merely through observing videotapes of their wild kin reacting to snakes. But when the tapes were spliced so that the wild monkeys appeared to be terrified by flowers and rabbits, the naive monkeys did not vicariously learn to fear these objects.[47] Although primate studies have not examined fears of spiders or insects, it is clear that "prepared learning" underlies fearful responses to at least some creepy creatures.

While studies of humans support the existence of an evolved predisposition for quickly learning to fear snakes and spiders, researchers also have probed the other half of learning. When it comes to fear of a dangerous object, an organism must both rapidly acquire the aversion and retain the lesson. When people were subjected to Pavlovian conditioning in which various pictures were paired with mild electrical shocks, participants showed stronger and more long-lasting responses to images of spiders than to images of houses, flowers, or mushrooms.[48] Through evolution, the learned association between spiders and pain is both readily acquired and "resistant to extinction." We don't soon forget a negative experience with a spider.

STINGING (AND FLUTTERING) REBUKES OF EVOLUTIONARY PSYCHOLOGY

As intriguing as the notion of adaptive fear might be, this evolutionary explanation for our negative response to insects has a few holes. First, when it comes to little creatures, we fear both dangerous and harmless (even downright beneficial) species.[49] Consider the moths. Although a few caterpillars have hairs that can elicit severe allergic reactions, the risk of these creatures to our survival is vanishingly small. Moreover, adult moths have never been a risk to human survival. But fear of moths is common enough to warrant its own term: mottephobia.

And what about my response to grasshoppers? While these insects have inflicted devastating losses to agriculturists in the past few thousand years, a swarm of locusts was a nutritional windfall for a much longer period of human history during which insects *were* our food.[50] From an evolutionary perspective, I should have been like a kid in a candy store—not a panicked adult—in that grasshopper-infested draw.

At least I take a more sane approach to spiders in my house, knowing that these eight-legged predators are keeping earwigs, flies, and their ilk at bay. However, most of us fear spiders despite their ecological value. And few people have an affection for bees, even though we've been satisfying our sweet tooth at their expense for millennia. In the grand scheme, it might well be argued that insects have been a net benefit to humanity,[51] with the notion of them as pests coming in the past ten thousand years as we crammed ourselves into cities where insect-vectored diseases such as bubonic plague became devastating and as we concentrated our food into homogenized production and storage systems where weevils and whatnot took their toll.

In response to this objection, the evolutionary psychologist might well appeal to the low cost of false positive alarms and the high cost of false negatives.[52] That is to say, my panicking in the midst of a grasshopper onslaught had a much lower cost than my poking at one of the many black widows that

Figure 2.3
Our fear of insects may be rooted in experience with pathogen-carrying insects, although major epidemics of insect-borne diseases did not emerge until humans began living in dense communities some ten thousand years ago. This seventeenth-century painting by Nicolas Poussin portrays the misery of bubonic plague in Ashdod, a city that is now part of modern-day Israel (image by Web Gallery of Art through Wikimedia Commons).

built webs in abandoned rodent burrows nearby. However, the supposed trade-off in evolutionary economics that favors fear is not empirically well supported.

The next objection to the evolutionary account of fear is that modern humans have much higher rates of entomophobia and arachnophobia than we have irrational fears of lions, bears, or crocodiles.[53] Even if the odd mosquito passed along a pathogen or the occasional nest of hornets killed one of our ancestors, surely big predators were more dangerous. But children adore teddy bears while exhibiting no great affinity for rubber spiders. The evolutionist might retort that while we are psychologically poised to learn a fear of predators, modern humans simply have many more opportunities to crystallize our latent aversion to cockroaches. Kids don't run across many tigers on the way to school, while they may well encounter a cockroach in the locker room. But this rebuttal leads to the final objection.

If "prepared learning" is foundational to our fear of insects, then entomophobes should have had negative encounters with these creatures. However, typically less than half of these individuals recall such conditioning experiences—in one study, only 8 of 118 subjects recalled such a traumatic event.[54] In response, the evolutionary psychologist could interpret this as evidence for the strong view that humans are born scared and our innate fear doesn't require practice. But this seems a bit too convenient, if no matter how we come by our fear there is an adaptationist account.

Even the obdurate evolutionist has to admit that what evolution has provided, culture has exploited. We are born with particular eye and hair colors, genetic traits that are arguably rooted in the reproductive success of our ancestors. At the end of the Ice Age it seems that European males preferred blue-eyed blondes, who were rare in the population and hence valued mates. Today's European men respond more favorably to brunettes, and blondes may be on their way out.[55] The point is that cultures can both favor traits and alter genetically endowed features (e.g., with contact lenses and hair dye). Evolution is a matching of biological and psychological features to the environment—and for us, the environment includes society. So when it comes to humans, little of our behavior can be understood without regard to how culture shapes our beliefs, thoughts—and fears.

NOTES

1. Charles Darwin, *The Descent of Man* (1871; repr., New York: Dover, 2010), 23.
2. Ibid.
3. Aaron T. Beck, Gary Emery, and Ruth L. Greenberg, *Anxiety Disorders and Phobias: A Cognitive Perspective* (New York: Basic Books, 2005), 4.
4. David H. Barlow, *Anxiety and Its Disorders: The Nature and Treatment of Anxiety and Panic*, 2nd ed. (New York: Guilford Press, 2002), 180.

5. Beck et al., *Anxiety Disorders and Phobias*, 4.
6. Ricky L. Langley, "Animal-related fatalities in the United States—An update," *Wilderness and Environmental Medicine* 16 (2005): 67–74.
7. Ricky Lee Langley and James Lee Hunter, "Occupational fatalities due to animal-related events," *Wilderness and Environmental Medicine* 12 (2001): 168–74.
8. Ricky Lee Langley, Sheila Ann Higgins, and Kitty Brown Herrin, "Risk factors associated with fatal animal-vehicle collisions in the United States, 1995–2004," *Wilderness and Environmental Medicine* 17 (2006): 229–39.
9. Pam Belluck, "Study of black bears finds it's not the mamas that should be feared the most," *New York Times*, May 11, 2011; Ronald Bailey, "North America's most dangerous mammal," *Reason*, November 21, 2001, http://reason.com/archives/2001/11/21/north-americas-most-dangerous (accessed December 10, 2011); Jennifer Viegas, "Shark attacks, resulting human deaths on the rise," Discovery News, March 1, 2010, http://news.discovery.com/animals/shark-attacks-resulting-human-deaths-on-the-rise.html (accessed December 10, 2011).
10. "Malaria," World Health Organization, http://www.who.int/mediacentre/factsheets/fs094/en (accessed December 10, 2011).
11. Stephen T. Asma, *On Monsters: An Unnatural History of Our Worst Fears* (New York: Oxford University Press, 2009), 4.
12. Alan Macfarlane, *The Savage Wars of Peace: England, Japan and the Malthusian Trap* (1997; repr., New York: Palgrave Macmillan, 2003), 214.
13. Martin Antony and David H. Barlow, "Specific phobias," in Barlow, *Anxiety and Its Disorders*.
14. Ibid.
15. Barlow, *Anxiety and Its Disorders*, chap. 6.
16. Ibid.
17. Andrew J. Tomarken, Susan Mineka, and Michael Cook, "Fear-relevant selective associations and covariation bias," *Journal of Abnormal Psychology* 98 (1989): 381–94.
18. Arne Öhman and Joaquim J. F. Soares, "'Unconscious anxiety': Phobic responses to masked stimuli," *Journal of Abnormal Psychology* 103 (1994): 231–40.
19. Arne Öhman and Susan Mineka, "The malicious serpent: Snakes as a prototypical stimulus for an evolved module of fear," *Current Directions in Psychological Science* 12 (2003): 7.
20. Vanessa Lobue and Judy S. Deloache, "What's so special about slithering serpents? Children and adults rapidly detect snakes based on their simple features," *Visual Cognition* 19 (2011): 129–43.
21. Stephen R. Kellert, "Values and perceptions," *Cultural Entomology Digest*, no. 1 (1993), Insects.org, http://www.insects.org/ced1/val_perc.html (accessed January 15, 2011).
22. Phillip Weinstein, "Insects in psychiatry," *Cultural Entomology Digest*, no. 2 (1994), Insects.org, http://www.insects.org/ced2/insects_psych.html (accessed October 31, 2011).
23. Jamie Bennett-Levy and Theresa Marteau, "Fear of animals: What is prepared?," *British Journal of Psychology* 75 (1984): 37–42.
24. Eric C. Brown, "Reading the insect," in *Insect Poetics*, ed. Eric C. Brown (Minneapolis: University of Minnesota Press, 2006), ix.
25. Maurice Maeterlinck, in *Hearst's Magazine*, March 1920, quoted in Charlotte Sleigh, "Inside out: The unsettling nature of insects," in Brown, *Insect Poetics*, 289–90.
26. Marion Copeland, *Cockroach* (London: Reaktion Books, 2003), 81.

27. Joanne E. Lauck, *The Voice of the Infinite in the Small: Revisioning the Insect-Human Connection* (Mill Spring, NC: Swan, Raven, 1998).
28. Sigmund Freud, "From the history of an infantile neurosis," in *The Standard Edition of the Complete Psychological Works of Sigmund Freud*, ed. and trans. James Strachey (London: Hogarth Press, 1957), 90.
29. Ibid.
30. Nicky Coutts, "Portraits of the nonhuman: Visualizations of the malevolent insect," in Brown, *Insect Poetics*.
31. Weinstein, "Insects in psychiatry."
32. Sleigh, "Inside out," 283.
33. Sigmund Freud, "New introductory lectures on psycho-analysis," in *The Standard Edition*, 31–56.
34. Carl Jung, "Basic postulates of analytical psychology," in *The Collected Works of C. G. Jung*, ed. and trans. R. F. C. Hull (New York: Pantheon, 1960), 349.
35. Carl Jung, "Synchronicity: An acausal connecting principle," in *The Collected Works*.
36. Carl Jung, "The significance of the father in the destiny of the individual," in *The Collected Works*.
37. Eve Adamson and Gayle Williamson, *The Complete Idiot's Guide: Dream Dictionary* (New York: Alpha Books, 2007); Jo Jean Boushahla, *The Dream Dictionary: The Key to Your Unconscious* (New York: Pilgrim Press, 1983); Silvana Amar, *The Bedside Dream Dictionary* (New York: Skyhorse, 2007); Penney Peirce, *Dream Dictionary for Dummies* (Hoboken, NJ: Wiley, 2008).
38. Hugh Raffles, *Insectopedia* (New York: Pantheon, 2010), 202–3.
39. James Hillman, "Going bugs," *Spring: A Journal of Archetype and Culture* (1988): 71.
40. Ibid., 40–72.
41. Graham C. L. Davey, Angus S. McDonald, Uma Hirisave, G. G.Prabhu, Saburo Iwawaki, Ching Im Jim, Harald Merckelbach, Peter J. de Jong, Patrick W. L. Leung, and Bradley C. Reimann, "A cross-cultural study of animal fears," *Behaviour Research and Therapy* 36 (1998): 735–50.
42. Noam Chomsky, *Aspects of the Theory of Syntax* (Cambridge: MIT Press, 1969); John Rawls, *A Theory of Justice* (1971; repr., Cambridge, MA: Belknap Press of Harvard University Press, 2005); Jerry Fodor, *The Modularity of the Mind* (Aarhus, Denmark: Bradford, 1983); John Mikhail, *Elements of Moral Cognition: Rawls' Linguistic Analogy and the Cognitive Science of Moral and Legal Judgment* (New York: Cambridge University Press, 2011).
43. Öhman and Mineka, "The malicious serpent."
44. Barlow, *Anxiety and Its Disorders*, chap. 7.
45. Ibid., chap. 6.
46. Antony and Barlow, "Specific phobias."
47. Michael Cook and Susan Mineka, "Selective associations in the observational conditioning of fear in rhesus monkeys," *Journal of Experimental Psychology: Animal Behavior Processes* 16 (1990): 372–89.
48. Arne Öhman and Susan Mineka, "Fear, phobias and preparedness: Toward an evolved module of fear and fear learning," *Psychological Review* 108 (2001): 483–522.
49. Davey et al., "A cross-cultural study of animal fears."
50. Julieta Ramos-Elorduy, "Anthropo-entomophagy: Cultures, evolution and sustainability," *Entomological Research* 39 (2009): 271–88.

51. Stephen R. Kellert, *Kinship to Mastery: Biophilia in Human Evolution and Development* (Washington, DC: Island Press, 1997).
52. Barlow, *Anxiety and Its Disorders*, chap. 7.
53. Jacqueline Ware, Kumud Jain, Ian Burgess, and Graham C. L. Davey, "Disease-avoidance model: Factor analysis of common animal fears," *Behaviour Research and Therapy* 32 (1994): 57–63.
54. Graham C. L. Davey, "Characteristics of individuals with fear of spiders," *Anxiety Research* 4 (1992): 299–314.
55. Arifa Akbar, "How women evolved blond hair to win cavemen's hearts," *Independent*, February 27, 2006, http://www.independent.co.uk/news/science/how-women-evolved-blond-hair-to-win-cavemens-hearts-467901.html (accessed January 20, 2012).

CHAPTER 3

Learning to Fear: Little Miss Muffet's Lesson

ENTOMOPHOBIC PRACTICE MAKES PERFECT

Sometimes well-meaning scientists can be a bit dense. Dewey Caron has a doctorate in entomology from Cornell, and he's forgotten more about bees than I'll ever know. Caron wrote an analysis of entomophobia for the *American Bee Journal* in which he suggested that beekeepers need to educate the public: "The consensus is that since insect fears and apprehension of stinging insects is a learned response it can be unlearned. . . . we need to show them there is nothing to be afraid of." Makes good sense. But the article then recounts the story of a fellow in Africa who was attacked by a swarm of bees:

> The unfortunate man jumped into the shallow river as the bees literally coated his body. . . . he began to sicken from the effects of the venom. Vomiting, he managed to move into deeper water. . . . His head ached badly. He suffered from diarrhea so intense that he was incontinent.[1]

So much for education. If you weren't freaked out before, Dr. Caron's tale provides outstanding fodder for any latent entomophobia you might be harboring. When I was a kid, a bee flew down the back of my shirt and stung me. I retain a vivid memory of the event, but it doesn't compare to the fellow who spent four and a half hours in the river and received 2,243 stings.

Most of my youthful encounters with insects were less traumatic. When I was ten we moved to the outskirts of Albuquerque, where there were grasshoppers aplenty seeking the greenness of our yard. On lazy summer afternoons I'd snag a few and feed them to the black widows that colonized the cinder-block wall in the backyard. I remember being darkly enchanted by the spiders' lethal tactics. Maybe that day on the prairie decades later evoked a childhood sense of being entangled and unable to escape—or perhaps the insects crawling into my clothing took me back to the bee inside my shirt.

I don't know what memories conspired to induce my panic, but psychologists contend that adult reactions often reflect childhood learning. Figuring out how a person came to be fearful would seem to be a simple matter of asking, but various studies have found that many phobic people report that they have always had their fears.[2] Perhaps these people are evolutionarily hardwired, but it is possible that they have simply forgotten earlier events. A subjective sense of having always been afraid might be expected given that the average age of onset for arachnophobia is 4.7 years and that early trauma can result in childhood amnesia.[3]

Researchers have converged on three mechanisms through which early experiences can catalyze phobias: direct experience (e.g., a cockroach runs up a kid's pant leg), modeling (e.g., a kid sees his mother scream in terror at cockroaches), and instruction (a kid's father tells her a story about cockroaches burrowing into children's ears).

Of these routes, experiences that condition an individual to fear spiders or insects appear to be the most common, with 40 to 50 percent of subjects recalling a frightening event.[4] Some studies report little difference in the frequencies with which normal and arachnophobic individuals remember encounters with spiders, but the intensity of encounters—rather than their mere occurrence—may be the critical factor.[5] Given a traumatic event, fear can become entrenched through "false alarms" (the discomfort induced by harmless encounters) and parental feedback. Conversely, previous uneventful and positive interactions with spiders psychologically "immunize" children so they are less likely to be conditioned by aversive incidents.[6]

Adults can both reinforce and initiate fear in children through modeling. About 20 percent of children fearful of spiders and insects report learning their aversion from parents, although friends and siblings were also implicated.[7] Some research even suggests that such vicarious or observational learning can be more important than direct experience. In his review of entomophobia, Tad Hardy noted, "For centuries it was believed that expectant mothers frightened by an insect (or any other experience) passed their fear to the developing child."[8] The medieval explanation was closer than one might think. Although no in utero learning takes place, spider fears are commonly transmitted from mother to daughter by same-sex modeling. Fathers serve as models for less than 5 percent of phobics.[9] Of course, observational learning has a bright side—adults who model positive responses to insects can prevent the incipient fears of children from reaching phobic proportions.[10]

Instructional learning probably is the least frequent means of acquiring entomophobia, but it can be potent. Even inadvertent lessons can generate phobias, as in the case of a young girl who became terrified of insects after she was told that her sister, who contracted pneumonia, had died from a "bug."[11]

And as one might expect, stories and conversations emphasizing the favorable characteristics of insects can block the reification of fears. However, arachnophobes often avoid information about spiders, so there may be little chance for instructional learning to diminish their fear.[12] Moreover, while books, television programs, and movies can offer information that may mitigate entomophobia, most social messages exacerbate fear.

THE SIX GREAT FEARS OF SIX-LEGGED CREATURES

Entomophobia is rooted in six "fear-evoking perceptual properties."[13] Insects can: (1) invade our homes and bodies; (2) evade us through quick, unpredictable movements, to which it might be added that the furtive skittering of a cockroach, for example, with its head lowered as if slinking out of the room, evokes a sense that the creature is guilty or ashamed; (3) undergo rapid population growth and reach staggeringly large numbers, threatening our sense of individuality; (4) harm us both directly (biting and stinging) and indirectly (transmitting disease as well as destroying woodwork, carpets, book bindings, electrical wiring, and food stores); (5) instill a disturbing sense of otherness with their alien bodies—they are real-world monsters associated with madness (e.g., "going bugs");[14] and (6) defy our will and control through a kind of radical mindless or amoral autonomy.

A particularly compelling formulation of these frightening qualities of insects is provided by the Yale anthropologist Hugh Raffles (in the following, his lines are rearranged to align with the order of the six perceptual properties presented above):

> There is the nightmare of foreign bodies in our ears and our eyes and under the surface of our skin. . . . There is the nightmare of turning the overhead light on just as the carpet scatters. . . . There is the nightmare of fecundity and the nightmare of multitude. . . . There is the nightmare of their being out to get us. . . . There is the nightmare of too many limbs [and] the nightmare of awkward flight and the nightmare of clattering wings. . . . There is the nightmare of beings without reason.[15]

Western culture provides abundant opportunities for children and adults to learn that insects invade, evade, overwhelm, attack, perturb, and defy—and modern media are adept at tapping into these capacities. For example, these elements have been woven into several popular television series, including *Billy the Exterminator* (A&E), *Infested!* (Animal Planet), *Dirty Jobs* (Discovery), and *Fear Factor* (NBC)—which might have been better titled *Disgust Dilemma*. And the capacity of grasshoppers to evoke fear (see, it's not just me) is captured in Guy Smith's novel *Locusts*. This book reads like the screenplay for a B-movie (with notable similarities to the 2005 cinematic flop *Locusts: Day of Destruction*), with the author managing to push all of our fear buttons:

> The storm clouds were here already, speeding in at an alarming rate in a wide black line, blotting out the overhead midday sun. . . . Locusts! By the millions [*overwhelm*]. . . . On and on came the locusts, their appetites whetted by a feed of barley. Their hunger was at full pitch [*attack*]. . . . Mrs. Hatherton could not breathe. Locusts were jammed solidly in her nostrils and throat [*invade*]. . . . Those horrible eyes, fixed on [Mr. Hatherton] with expressions of hate, the repulsive bodies heaving as they breathed [*perturb*]. . . . These horrors had wings [*evade*]. . . . they were on him again in even greater numbers, raking, spitting and stinging, taking their revenge [*defy*].[16]

Before we explore how Western culture inculcates each of these half dozen fears into children and adults, it is important to consider that there are some positive images. My kids adored Eric Carle's books, for example, with their endearingly imperfect characters including a busy spider, a courageous honeybee, a quiet cricket, a clumsy beetle, a hungry caterpillar, and a grouchy ladybug.[17] Such stories build on an ancient tradition in which myths portrayed insects favorably: ants exemplified industry and thrift, bees evoked cooperation and chastity, butterflies represented beauty and transformation, crickets radiated happiness and domesticity, ladybugs provided good works and fortune telling, moths embodied the soul and afterlife, and scarab beetles symbolized renewal and creation.[18]

Outside the West, many indigenous peoples include insects among their totems. Clans of Australian Aborigines assure the protection of their namesakes, which represent vital food sources, such as edible beetle larvae, witchetty grubs, and honey ants.[19] And in the United States, athletic teams have adopted insect mascots, such as wasps, hornets, and scorpions, along with the less expected mosquitoes and boll weevils.[20]

While positive portrayals of insects and their kin can be found, for every rendition of "The Itsy Bitsy Spider" and every screening of *Microcosmos* there are a dozen rounds of "Little Miss Muffet" and showings of *The Fly* (the 1958 version or its 1986 remake, not to mention the opera). Western culture—and particularly the movies, which so powerfully reflect and shape social norms—reminds us that insects can . . .

Invade our homes, bodies, and minds

Insects were the featured invaders of the big screen in the 1950s, with giant ants (*Them!*, 1954), spiders (*Tarantula*, 1955), grasshoppers (*Beginning of the End*, 1957—I *knew* it), and even a praying mantis (*Deadly Mantis*, 1957). The

Figure 3.1
This 1940 poster from the Works Progress Administration portrays the classic story of the spider and Miss Muffet in less negative terms than we might expect in an effort to promote reading among children. But even this encounter seems to have evoked surprise—and perhaps fright—judging by the eyes of the cartoon girl (image by Gregg Arlington of the WPA Federal Arts Project).

conventional view is that these creatures were Cold War metaphors for our anxieties about communists taking over the world, along with our worries about technology (radiation invariably triggered the insectan incursions).[21] However, at least one historian contends that sometimes a mantis is just a mantis and that these films simply reflected our fear of insects invading our homes.[22] If so, this might explain Hollywood's continuing to produce films featuring insects insinuating themselves into our lives even after the collapse of the Soviet Union (e.g., *Mimic*, 1997, which was followed by *Mimic 2*, 2001, and *Mimic 3: Sentinel*, 2003; *Spiders*, 2000, and the sequel *Spiders 2: Breeding*

Ground, 2001; and *Tail Sting*, 2001—a predecessor of *Snakes on a Plane*, 2006, but with bioengineered scorpions). Even the relatively staid National Geographic Channel has exploited the entomophobic proclivities of viewers with an episode of *Nature's Nightmares* featuring insectan home invasions.

Insects are even scarier when they enter our bodies. "I know an old lady who swallowed a fly"—and from there every kid knows how things spin out of control. The lesson: Don't let insects into your body or perhaps you'll die. And when you die: "The worms crawl in, the worms crawl out / The worms play pinochle on your snout." The old wives' tale of earwigs is vividly exploited in *Star Trek: The Wrath of Kahn* (1982) when the villain places what look like alien antlion larvae in the ears of his prisoners, where they enter the victims' brains, rendering them "extremely susceptible to suggestion."

Bodily invasion is the name of the game on *Fear Factor*, a show in which contestants put live insects (e.g., beetles, crickets, and cockroaches) in their mouths and either consume them or transfer the contents to a partner's mouth. In other episodes, shapely contestants wearing little clothing are covered in lots of insects to suggest the titillating possibility of bodily invasion. The British version of the program was censored by the Broadcasting Standards Commission, which decreed that "graphic and extended footage of the consumption and treatment of the insects, purely for entertainment, had exceeded acceptable boundaries."[23] It's hard to keep a stiff upper lip with a mouthful of maggots.

Once insects get into our bodies, it's a small step to infesting our minds. In the original and the remake of *The Fly*, the hero's body is melded with that of a fly when the insect inadvertently enters a teleportation machine as the scientist is testing the device on himself. What begins as an anatomical invasion becomes increasingly psychological as the chimeric character begins to think and act with the amoral tendencies of an insect. In the remake, the fellow becomes a sexual dynamo and impregnates his girlfriend, who, when she figures out that he's slowly metamorphosing into a fly, seeks an abortion, declaring, "I don't want it in my body."

But if the arthropod insinuates itself only into our bodies, we might get the best of all worlds. In the Spider-Man comics and movies, the hero is bitten by a radioactive spider and subsequently acquires the powers of an arachnid. However, Peter Parker does not take on the psychological attributes of the spider. Rather than becoming a cold-hearted predator, he retains his human and humane mind and becomes a superhero. But not all (or most) transformations go so smoothly.

In *The Wasp Woman* (1959), an aging cosmetic mogul takes "wasp enzymes" to rejuvenate her body, and along with gaining youthfulness she periodically transforms into a bug-eyed homicidal maniac. And taking a page from *Fear Factor*, characters in *Starship Troopers 2* (2004) pass their insectan parasites from mouth to mouth. The new host is converted into a zombielike minion devoted to spreading the infestation to others—the horror genre's version of being plagued by an "earworm," which is both the common name of a moth larva that infests corn and the term for that maddening condition of having a song stuck in your head.

The most terrifying cinematic portrayal of an infested mind is found in *Bug* (2006), based on the play by Tracy Letts. In the film, a honky-tonk waitress shacks up with a drifter who suffers from the delusion that he's infested with insects, and she is inexorably drawn into his madness. At first he simply feels a tiny insect bite him in their bed, next he's spraying the apartment, then she feels something under her skin, and then he's yanking out a tooth to get to the egg sac that the government implanted. By the final scene, they have draped their apartment with mosquito netting, hung dozens of bug zappers, and slashed their bodies in a futile effort to extract the imaginary insects. As the authorities arrive to end the insanity, the couple strip, splash gasoline over themselves, babble insanely about bugs, declare their love, and strike a match. So it goes when insects invade—and when they . . .

Evade our efforts to detect and destroy them

Insects are devious. Even giant insects are masters of concealment. In the cult classic *Them!* mutant ants manage to elude their human pursuers in caves, sewer systems, and cargo ships. And when not hiding, insects in the movies are sneaking. Spiders are notorious for their scurrying, as in the intentionally funny *Arachnophobia* (1990) and the unintentionally funny *Spiders* (and *Spiders 2: Breeding Ground*). At least the evil ones behave this way. Interestingly, when insects are portrayed as good (e.g., Jiminy Cricket in *Pinocchio*, 1940, and Flik, the hero ant in *A Bug's Life*, 1998) they exchange six-legged skittering for bipedal striding.

Nor are the evasive tactics of hiding and fleeing mutually exclusive. For example, in *Ticks* (1993) and *Bugged* (1996, a horror movie wherein the only real horror is that the film was ever made), the chemically catalyzed foot-long arthropods are extraordinarily secretive, and swift. Likewise in a nightmarish scene from the *Billy the Exterminator* episode featuring cockroaches, a woman screams in horror as a pile of cockroaches previously hidden beneath a trash can scatters in all directions. And this leads us to the realization that insects . . .

Overwhelm our sense of individuality

As the cockroaches scurry, Billy observes, "We've got literally five hundred to a thousand at this one location. . . . They've taken over." In a later episode he extracts five thousand bees from a hive they've built between the walls of a house. But these numbers pale in comparison to the goosebumps raised by *Dirty Jobs* host Mike Rowe's encounter with the twenty-five million insects seething in bins and vats at Ghann's Cricket Farm.

Throughout Western literature, masses of insects have played a disturbing role; they have been described as "horrors of nonindividual groups [in which] the power of swarms was a frightening one, emerging from the sheer size of the pack . . . as in the case of locusts."[24] Indeed. One of the movies' most fantastic challenges to our precious individuality unfolds in the 2008 remake of *The Day the Earth Stood Still*. In this film, aliens release tiny, fast-reproducing bugs that coalesce into an enormous, machine-like humanoid that consumes every form of matter and energy—all to save humanity from its own technological and moral follies. According to film critic Matt Mueller, the monster's form and power are based on those of a locust swarm.[25]

Other insects also have played the role of numerical villains, as with a band of Amazonian army ants constituting a "monster twenty miles long and two miles wide . . . forty square miles of agonizing death!" (*The Naked Jungle*, 1954). Killer bees wreak havoc in at least five films released in the 1970s (e.g., *The Swarm*, 1978).[26] And thanks to insecticides wiping out their food supply, tarantulas form "an army of deadly predators" to overwhelm larger prey, including livestock, in *Kingdom of the Spiders* (1977).

Being overwhelmed by insects can also be a matter of monstrous size. Just as an angry God sent locust swarms to punish an arrogant Pharaoh, the giant insects of twentieth-century films were the consequences of abundant hubris and deficient foresight. During the Cold War, radiation generated school-bus-sized carnivorous locusts (*Beginning of the End*), ten-foot ants (*Them!*), and monstrous moths (*Mothra*, 1961). By the 1970s, pollution was the sin, and humans were punished by giant ants spawned from a toxic waste dump (*Empire of the Ants*, 1977).[27] A misguided geneticist's attempt to play God created a vampire moth—a woman who transformed at night into a giant, blood-sucking moth in *The Blood Beast Terror* (1968). And not having learned our lesson, twenty years later we saw genetic engineering produce killer cockroaches in *The Nest* (1988).

While huge numbers (and sizes) of insects assail our modern sensibilities, for centuries Western culture has reminded us that these creatures . . .

Attack our health, food, and property

If the Bible is our guide to faith and fear, then it's little wonder that we're entomophobic. Tallying up references to insects in the Bible reveals forty-six negative allusions (e.g., "At his command came swarms of flies and maggots the whole land through"; Psalms 105:31) and just four positive mentions (e.g.,

"Go to the ant, you sluggard, watch her ways and get wisdom"; Proverbs 6:6).[28] On the Godly side, Yahweh coerces a recalcitrant Pharaoh with gnats, flies, and locusts. On the demonic side, Beelzebub, the prince of the devils, is the "lord of the flies." And in the Middle Ages, dragonflies were called the devil's darning needles and said to be giant flies sent by Satan to sew up the mouths of lying children.[29]

With this cultural momentum, it is not surprising that insects continue to be cast as villains. Eric Carle has an uphill battle in children's literature given works such as Douglas Florian's *Insectlopedia*.[30] Dangerous insects in the book include army ants ("You're lucky if / We miss your place."), dragonflies ("The demon of the skies . . . I terrorize"), locusts ("Your grain, / Your grass. / They disappear / Each time we pass"), and mosquitoes ("They feast on your skin / For take-out food."). Although the illustrations are whimsical, the poems can be dark:

> We are weevils.
> We are evil.
> We're aggrieved.
> Since time primeval.
> With our down-curved
> Beaks we bore.
> Into crops
> And trees we gore.
> We are ruinous.
> We are rotten.
> We drill holes
> In bolls of cotton.
> We're not modern,
> We're medieval.
> We are weevils.
> We are evil.[31]

The biblical framing of insects has been woven into painting and literature throughout the history of Western culture,[32] and it persists today in even low-brow cinematic ventures. For example, halfway into *Locusts: Day of Destruction*, a US senator is briefed on the swarms of bioengineered locusts ravaging the nation, and he compares the unfolding disaster to the Apocalypse. Indeed, the movie portrays locusts in terms of the Four Horsemen: pestilence, war, famine, and death.

As for pestilence, *Locusts* begins with scientists modifying the Australian *plague* locust into a "bioweapon" (a macroscopic version of germ warfare); the experts later advocate the imposition of a national quarantine to keep the infestation from becoming a pandemic. The element of war emerges as the US

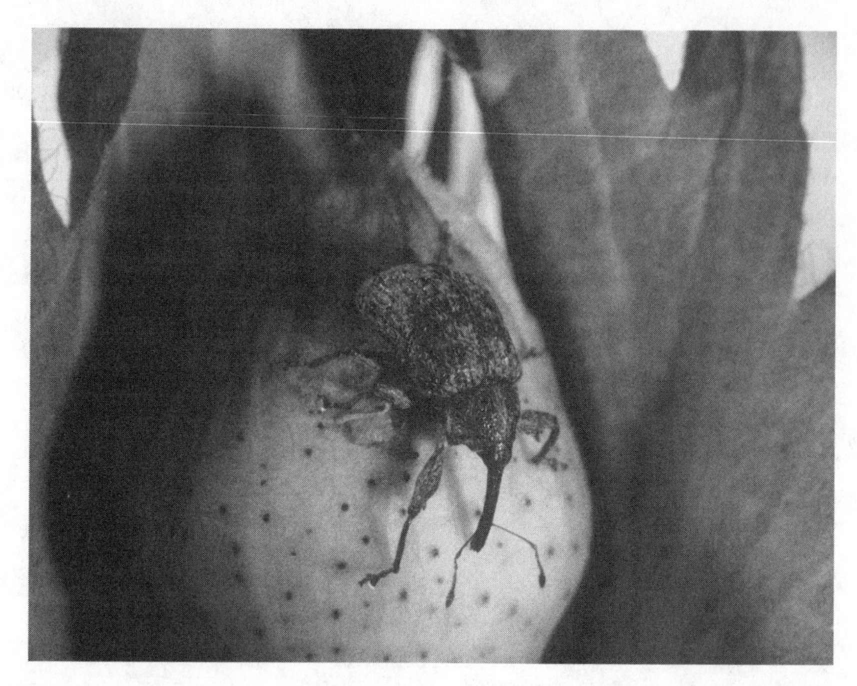

Figure 3.2
The boll weevil has been an object of considerable cultural attention, including a monument in Enterprise, Alabama, crediting the insect with forcing farmers to diversify their crops, which led to prosperity. However, most allusions in song and story, including "Boll Weevil Blues" and Douglas Florian's poem for children, portray the weevil as devastating (image by US Department of Agriculture).

military prepares to spray the American heartland with nerve gas, as it is the only chemical lethal to the locusts. Thousands of unsuspecting rural folks will die if the heroine cannot come up with an alternative plan. Such staggering collateral casualties are justified in light of the third horseman, for unless the locusts are stopped there will be worldwide famine. As farmers race to harvest their crops, the fourth horseman arrives in various guises. Death comes in the forms of a massive traffic pile-up and an airplane crashing in a fiery ball after locusts clog the engines. After a swarm attacks livestock, the government expert warns, "Worst case scenario, they turn carnivorous," to which the senator ominously responds, "We are on the menu." A scene with a blood-soaked, locust-ravaged victim reveals the fate of those who are not gassed or starved. Humanity is saved, however, when the heroine turns the nation's power grid into the world's largest bug zapper.

This movie fails to horrify, as inept writing, acting, and directing make it a presumably unintentional parody of one of the most chilling films in cinematic history. *The Birds* (1963) is the one movie of my childhood that I cannot get out of my mind. Perhaps Alfred Hitchcock planted a seed—the sense of being trapped in a swarm of creatures that are weirdly both mindless and malevolent—that germinated in a grasshopper-filled draw years later. And this takes us to the realization that insects . . .

Perturb our tranquility through their monstrous otherness

As alien beings, insects have enabled painters, dancers, essayists, novelists, poets, filmmakers, and advertisers to evoke fear for centuries.[33] In the Dark Ages, the Church maintained that insects were spawned from pockets of sin within cadavers, a macabre transformation that was captured in paintings such as Matthias Grünewald's *Dead Lovers* (ca. 1480) and continued through William Blake's insect-human hybrid *The Ghost of a Flea* (1812).

While such visual art is unsettling, the mind's eye can see into even darker places. And no piece of literature taps into monstrous otherness more powerfully than Franz Kafka's *The Metamorphosis*, in which Gregor Samsa wakes up to find himself transformed into a gigantic insect (a cockroach being the favored interpretation). Even those who once loved him are repulsed:

> [Gregor's mother] looked first with hands clasped together at his father, then took two steps towards Gregor and collapsed, surrounded by her outspread skirts, her face sunk and quite hidden in her breast. His father clenched his fist with a hostile expression, as if meaning to drive Gregor back into his room.[34]

Kafka may not have originated the existential fear of becoming an insect, but his influence on Western art and literature has persisted. His disturbing touch can be felt in *The Fly*, *The Wasp Woman*, *The Blood Beast Terror*, and *Starship Troopers 2*. However, the most terrible transformations of humans into insects and monsters have taken place not in art, but in politics.

Given the human capacity for empathy, genocide requires an extraordinary alteration of the human psyche. Careful analysis has revealed an eight-stage process, the first three steps of which involve cognitive restructuring in which the enemy is classified, symbolized, and dehumanized.[35] This depraved conceptual metamorphosis often involves turning people into vermin:

> We must kill them big and little. . . . Nits make lice.
>
> —Colonel John M. Chivington, exhorting his troops to massacre a Cheyenne community, including women and children, in 1864[36]

> Antisemitism is exactly the same as delousing. Getting rid of lice is not a question of ideology. It is a matter of cleanliness.
>
> —Heinrich Himmler, April 24, 1943[37] (Adolf Hitler was profoundly entomophobic and framed the Holocaust in terms of extermination.)

> If you gave me a pesticide to throw at these swarms of insects [the Iranians] to make them breathe and become exterminated, I would use it.
>
> —Iraqi general Maher Abd al-Rashid, 1984[38]

> The Inyenzi [cockroaches] have always been Tutsi. We will exterminate them.
>
> —Broadcast from a Hutu-run radio station during the Rwandan genocide, 1994[39]

> It should be legal to shoot [Mexicans] on sight. They breed their filthy race like the cockroaches that they are.
>
> —Post on Rush Limbaugh Campfire website, 2002[40]

> Bugsplat is the official term used by US authorities when humans are killed by drone missiles.
>
> —Jennifer Robinson, human rights lawyer, 2011[41]

If turning humans into insects countenances hate, then turning insects into humans has the opposite effect. Artists humanize insect heroes by transforming their alien features into eyes, mouths, heads, and appendages more like our own. In *Antz* (1998), the "good" ants stand upright and their forelegs function as arms with human-like fingers, while the "bad" termites run around on six legs. Likewise, in *A Bug's Life* the "good" ants are anthropomorphized into smooth-bodied beings with two arms, two legs, and human-like teeth. The "bad" grasshoppers have spiky bodies, six appendages (they are bipedal but have four "arms"), and jagged mandibles.

Animators are also keenly aware that we find infantile features charming, so "good" insects are drawn with proportions similar to those of a human baby. Infants have disproportionately large heads; their noggins are about 50 percent of the length of their bodies, versus 20 percent for adults. Disney's Jiminy Cricket in *Pinocchio* is phenomenally infantile by this measure, with a head that is 70 percent of his body length (versus about 15 percent in an actual cricket)! The eyes of a human baby are also enlarged, being about twice the relative size of adults' eyes. So to soften the insectan features of the hero of *A Bug's Life*, the animators gave him eyes that are a remarkable 50 percent of his head length (about twice the proportion in an actual ant). But even when we put a humanoid head on an ant, insects can still . . .

Defy our will and control through their amoral autonomy

In some social-historical contexts, insects have served as positive representations of determination and hard work. The Mormons who settled Utah used the beehive as an emblem of their own industriousness, and the imagery has lived on in the Beehive State.[42] However, the tension between the authoritative control in a hive and the autonomy of the workers has played out in a rather strange way in Utah as an extremely conservative state government has employed strikingly socialist policies.[43] And the oppressive implications of aligning with the social insects have come to dominate our imagery.

With the rise of industry, insects began to serve as sinister metaphors for powerful, efficient, cold, dangerous, heartless machines. This theme continued into the twentieth century, during which "the precision of the insect machine fed into dystopic fears of 'inhuman technological society' modeled on emotionless insect minds."[44] This fear was captured in such works as Karel Čapek's *The Insect Play* (1921) and the film *Metropolis* (1927), featuring a queen-like woman commanding servile, ant-like workers. And with the emergence of totalitarian governments, insects came to represent not only the living cogs of industrialism but also the submissive followers of despotism.

In *Poetics of the Hive*, Cristopher Hollingsworth traces the ways in which insects have been used in literature and film to frame social issues and political fears.[45] With the rise of fascism and communism in the twentieth century, the lives of ants, termites, and bees became ominous representations of forced collectives in which individuals were enslaved and sacrificed for the good of

Figure 3.3
Although social insects such as bees were seen in largely virtuous terms prior to the twentieth century, with the rise of totalitarian regimes these creatures (along with ants and termites) became associated in film and literature with the sacrifice of the individual for the collective, the plight of servile factory workers, and the oppression by centralized governments (image by Vagawi through Creative Commons).

the whole. In the 1920s, Maurice Maeterlinck explored the staggering costs of perfect communal order in *The Life of the White Ant.*[46] And almost a century later, the human-turned-insect in *Starship Troopers 2* declared that humanity must be destroyed because it glorifies "the one over the many."

Even children's books and movies present insects as soulless automata. In the 1912 book *The Adventures of Maya the Bee* the main character sacrifices herself for the good of her hive.[47] Both *A Bug's Life* and *Antz* portray insects as docile servants of those in power. Individuals who aspire to be more than dispassionate drones are violently oppressed, although they eventually rise above the brutal hierarchy.

Finally, the uncanny power of insects is manifest in dark, psychic powers. Maeterlinck's novels allude to the "spirit of the hive," with communication taking place not as conscious language but "as affects of murmur, whisper, and a refrain that even the bees might not hear but sense in some uncanny way."[48] Likewise, Sylvia Plath's 1962 poem "The Swarm" attributes "A black intractable mind" to insect masses.[49] And Virginia Woolf taps into an old superstition in her 1922 novel *Jacob's Room*, with a death's-head moth (named for the skull-shaped marking on the thorax) serving as an eerie harbinger of death.[50]

THE DISGUSTING ELEPHANT IN THE FEARFUL ROOM

Scary stories, songs, poems, paintings, and films involving insects often have another emotional component. Gregor Samsa, the giant insect in Kafka's *Metamorphosis*, is not only or even primarily frightening. Rather, he is detestable and elicits disgust. And the most appalling scene in the remake of *The Fly* involves the insectan fellow vomiting a white slime onto his food to begin digestion. This same sense of filth is vividly captured by Mark Twain:

> The fly . . . hunts up patients suffering from loathsome and deadly diseases; wades in their sores, gaums its legs with a million death-dealing germs; then comes to that healthy man's table and wipes these things off on the butter and discharges a bowel-load of typhoid germs and excrement on his batter cakes.[51]

A growing body of research suggests that fear and disgust are entangled. After all, these are the two universal aversive emotions. Indeed, the best predictor of entomophobia appears to be parental disgust sensitivity, and arachnophobic girls tend to have mothers with a strong disgust of spiders.[52] It seems that when insects infest our minds—through evolution or enculturation—they bring with them not only the capacity to invade, evade, overwhelm, attack, perturb, and defy. They also bring an organic, viscid, ugly, soft, wriggling filthiness.

NOTES

1. Dewey M. Caron, "Entomophobia!," *American Bee Journal* 130 (1990): 782, 780.
2. Richard J. McNally and Gail S. Steketee, "The etiology and maintenance of severe animal phobias," *Behaviour Research and Therapy* 23 (1985): 431–35; Mairwen K. Jones and Ross G. Menzies, "The etiology of fear of spiders," *Anxiety, Stress, and Coping* 8 (1997): 227–34; Harald Merckelbach and Peter Muris, "The etiology of childhood spider phobia," *Behaviour Research and Therapy* 35 (1997): 1031–34.
3. Merckelbach and Muris, "The etiology of childhood spider phobia."
4. McNally and Steketee, "The etiology and maintenance of severe animal phobia"; Harald Merckelbach, Arnoud Arntz, and Peter J. de Jong, "Conditioning experiences in spider phobics," *Behaviour Research and Therapy* 29 (1991): 333–35; Harald Merckelbach, Arnoud Arntz, Willem A. Arrindell, and Peter J. de Jong, "Pathways to spider phobia," *Behaviour Research and Therapy* 30 (1992): 543–46; David H. Barlow, *Anxiety and Its Disorders: The Nature and Treatment of Anxiety and Panic*, 2nd ed. (New York: Guilford Press, 2002), chap. 7.
5. Merckelbach et al., "Pathways to spider phobia."
6. Merckelbach and Muris, "The etiology of childhood spider phobia."
7. McNally and Steketee, "The etiology and maintenance of severe animal phobia"; Merckelbach et al., "Conditioning experiences in spider phobics"; Peter Muris and Harald Merckelbach, "Treating spider phobia with eye-movement desensitization and reprocessing: Two case reports," *Journal of Anxiety Disorders* 9 (1995): 439–49.
8. Tad N. Hardy, "Entomophobia: The case for Miss Muffet," *Bulletin of the Entomological Society of America* 34 (1988): 65.
9. Merckelbach and Muris, "The etiology of childhood spider phobia"; Peter J. de Jong, Helene Andrea, and Peter Muris, "Spider phobia in children: Disgust and fear before and after treatment," *Behaviour Research and Therapy* 35 (1997): 559–62.
10. Edward P. Sarafino, *The Fears of Childhood: A Guide to Recognizing and Reducing Fearful States in Children* (New York: Human Sciences Press, 1986).
11. Harold Kolansky, "Treatment of a three-year-old girl's severe infantile neurosis," *Psychoanalytic Study of the Child* 15 (1960): 261–85.
12. Merckelbach et al., "Pathways to spider phobia."
13. Jamie Bennett-Levy and Theresa Marteau, "Fear of animals: What is prepared?," *British Journal of Psychology* 75 (1984): 37–42; David N. Byrne, Edwin H. Carpenter, Ellen M. Thoms, and Susanne T. Cotty, "Public attitudes toward urban arthropods," *Bulletin of the Entomological Society of America* 30 (1984): 40–44; Stephen R. Kellert, "The biological basis for human values of nature," in *The Biophilia Hypothesis*, ed. Stephen R. Kellert and Edward O. Wilson (Washington, DC: Island Press, 1993), 57–58. To package our revulsion more evocatively, consider Charlotte Sleigh's interpretation: "Insects are all wrong. There is a good case for regarding them as zoology's Other, the definitive organism of *différance*. We humans have skeletons; they keep their hard parts on the outside and their squishy bits in the middle. We humans celebrate intelligence as our defining feature; they form almost equally complex societies by instinct. No wonder we are disgusted and fascinated when we find them in the kitchen." Charlotte Sleigh, "Inside out: The unsettling nature of insects," in *Insect Poetics*, ed. Eric C. Brown (Minneapolis: University of Minnesota Press, 2006), 281.
14. James Hillman. "Going bugs," *Spring: A Journal of Archetype and Culture* (1988): 40–72.

15. Hugh Raffles, *Insectopedia* (New York: Pantheon, 2010), 202–3.
16. Guy N. Smith, *Locusts* (London: Sheridan, 1979), 112–16, 213–14.
17. Eric Carle's books include, *The Grouchy Ladybug* (New York: HarperCollins, 1977); *The Very Hungry Caterpillar* (New York: Philomel Books, 1983); *The Very Busy Spider* (New York: Philomel Books, 1984); *The Very Quiet Cricket* (New York: Philomel Books, 1990); and *The Very Clumsy Click Beetle* (New York: Philomel Books, 1999).
18. Gene Kritsky and Ron Cherry, *Insect Mythology* (New York: Writers Club Press, 2000).
19. May R. Berenbaum, *Bugs in the System: Insects and Their Impact on Human Affairs* (New York: Helix Books, 1995).
20. Ibid.
21. Richard J. Leskosky, "Size matters: Big bugs on the big screen," in Brown, *Insect Poetics*.
22. William M. Tsutsui, "Looking straight at *Them!* Understanding the big bug movies of the 1950s," *Environmental History* 12 (2007)237–53.
23. Mark Jeffries, "Insect eating to be banned: You've had your Phil of TV bugs," *Daily Star*, September 30, 2003.
24. Jussi Parikka, *Insect Media: An Archaeology of Animals and Technology* (Minneapolis: University of Minnesota Press, 2010), 50.
25. Matt Mueller, "Excellent adventure, or bogus journey?," *Total Film*, December 2008, 68–72.
26. Tsutsui, "Looking straight at *Them!*"
27. Leskosky, "Size matters."
28. Kritsky and Cherry, *Insect Mythology*.
29. Ibid.
30. Douglas Florian, *Insectlopedia* (New York: Voyager Harcourt, 1998).
31. Ibid., 25.
32. Kritsky and Cherry, *Insect Mythology*; Nicky Coutts, "Portraits of the nonhuman: Visualizations of the malevolent insect," in Brown, *Insect Poetics*; Yves Cambefort, "A sacred insect on the margins: Emblematic beetles in the Renaissance," in Brown, *Insect Poetics*.
33. Eric C. Brown, ed., *Insect Poetics* (Minneapolis: University of Minnesota Press, 2006).
34. Franz Kafka, *The Metamorphosis, and Other Stories* (New York: Oxford World's Classics, 2009), 39.
35. Cristopher Hollingsworth, "The force of the entomological other: Insects as instruments of intolerant thought and oppressive action," in Brown, *Insect Poetics*.
36. Quoted in Steven Mintz and Sara McNeil, "Tragedy of the Plains Indians," Digital History, http://www.digitalhistory.uh.edu/disp_textbook_print.cfm?smtid=2&psid=3499 (accessed February 20, 2012).
37. Quoted in Raffles, *Insectopedia*, 141.
38. Quoted in Efraim Karsh, *The Iran-Iraq War* (New York: Rosen, 2002), 51.
39. Quoted in Mary Kimani, "Media trial: 'Hate radio' urged Hutus to break 'small noses,' expert witness testifies," Internews, March 20, 2002, cited in Hollingsworth, "The force of the entomological other."
40. William, "Mexicans are human garbage," on Rush Limbaugh Campfire, March 6, 2002, http://killdevilhill.com/conservativechat/shakespearew/31.html (accessed February 20, 2012).
41. Jennifer Robinson, "'Bugsplat': The ugly US drone war in Pakistan," Al Jazeera, November 29, 2011, http://www.aljazeera.com/indepth/opinion/2011/11/201111278839153400.html (accessed February 20, 2012).

42. "Emblem and Motto," Utah.com, http://www.utah.com/visitor/state_facts/symbols.htm (accessed November 9, 2012).
43. Jonathan Thompson, "Red state rising: How the Mormon GOP runs Utah with a collectivist touch," *High Country News*, October 29, 2012, 14–26.
44. Parikka, *Insect Media*, 42.
45. Cristopher Hollingsworth, *Poetics of the Hive* (Iowa City: University of Iowa Press, 2001).
46. Maurice Maeterlinck, *The Life of the White Ant* (1927; repr., Whitefish, MT: Kessinger, 2005).
47. Waldemar Bonsels, *The Adventures of Maya the Bee* (1912; repr., Whitefish, MT: Kessinger, 2004).
48. Parikka, *Insect Media*, 50.
49. Sylvia Plath, "The Swarm," 1962, reprinted in *Guardian*, March 12, 2008, http://www.guardian.co.uk/books/2008/mar/13/poetry.sylviaplath1 (accessed February 20, 2012).
50. Virginia Woolf, *Jacob's Room* (1922; repr., New York: Melville House, 2011).
51. Mark Twain, *Letters from the Earth*, ed. Bernard DeVoto (Greenwich, CT: Fawcett Crest, 1964), 35.
52. Graham C. L. Davey, Lorna Forster, and George Mayhew, "Familial resemblances in disgust sensitivity and animal phobias," *Behaviour Research and Therapy* 31 (1993): 41–50; de Jong et al., "Spider phobia in children."

CHAPTER 4

A Fly in Our Mental Soup: How Insects Push Our Disgust Buttons

Within minutes, our hands were covered in feces and vomit. Our quarry was the plains lubber grasshopper, the largest of all insects on the Wyoming grasslands—and its appearance matches its disgusting behavior. *Brachystola magna* resembles the human archetype of repulsion: the old hag. This is a fat grasshopper with deep wrinkles and folds, grotesquely beefy femurs, a pathetically spare rear-end, and a bald head (at least it doesn't have hair growing in offensive places). This is no dainty grasshopper capable of lithesome leaps; it has the heft of a breakfast sausage. Every summer we collected a few dozen of these creatures for dissection in the "Insect Anatomy & Physiology" laboratory.

Most of my summer encounters with rangeland grasshoppers involved attempts to suppress grasshopper outbreaks. There is a certain nobility to taking on a worthy foe that humans have battled for centuries. However, there was no such glory in my encounters with lubber grasshoppers, which don't reach outbreak proportions and certainly can't swarm, given that the lubberly beasts have wings that are reduced to useless stubs. And if they threatened farmers' fields, mounting a control program would be tantamount to picking a fight with the fat kid on the playground or beating up the town drunk.

Moreover, the lubber is about the easiest species of grasshopper to catch, at least in principle. However, the only way to gather dozens of them is by hand. An insect net is an effective means of capturing grass*hoppers* that are willing and able to live up to their names, but a net will snare few of these lumbering creatures. They are clumsy behemoths, hopping with the agility of insectan sumo wrestlers. Hence the name lubbers. Grabbing a fat, flightless beast—the dodo of the insect world—is a simple matter. But catching this grasshopper is not the same thing as holding on to it.

Figure 4.1
Brachystola magna is an archetype of a disgusting insect. The grasshopper's common name, the plains lubber, refers to its big, clumsy, and off-putting appearance. When caught, this bald, wrinkled, obese, and flightless grasshopper struggles powerfully and both regurgitates a brown fluid and excretes mushy feces, which become smeared on the captor (image by Marco Zanola through Wikimedia Commons).

The lubber may look dim-witted and benign, but appearances can be deceiving. Lurking beneath the bulging exoskeleton is a cantankerous creature. One must be careful when accosting these grasshoppers because their hind legs sport rows of spines that they rake across the flesh of a would-be captor. A clever predator (or entomologist) can neutralize this defensive maneuver by grabbing them by their hind legs, but at this juncture, the lubbers resort to their most noteworthy tactic: they become utterly repulsive.

Their first and most revolting strategy in this regard is to regurgitate copiously. Many species exhibit this defensive behavior, and as kids we referred to grasshoppers as "spitting tobacco juice." Indeed, the cola-colored fluid resembles the expectorant of tobacco chewers in its capacity to stain whatever it hits. Of course, a grasshopper isn't spitting wads of chewed tobacco. Rather, it is heaving up masticated and liquefied sunflower leaves—the contents of its foregut, which is the anatomical equivalent of our stomach. The prairie lubber manages to produce this material in impressive quantities, smearing itself and its handler with the dark brown fluid. For this grasshopper, however, the effort to repulse an assailant is not complete.

The restrained grasshopper next begins to defecate prodigiously. As opposed to vomiting, this offensive approach is not widely practiced among the lubber's brethren. Perhaps it wouldn't be particularly repugnant for most species, as grasshoppers generally produce very dry, compact fecal pellets the size of sesame seeds. The lubber, in contrast, can produce a dozen mushy turds, similar to those of a mouse, in quick succession. Its favorite meal of juicy roadside sunflower leaves is far more succulent than prairie grasses and provides enough fluid to allow this grasshopper the luxury of a soft stool. Thus, an experienced collector avoids the rear end of the grasshopper and holds the insect at bay for a few seconds until it has exhausted its colonic arsenal. If one is too hasty in dropping the repulsive creatures into a collecting bag, the grasshoppers quickly foul the container with smeared feces, making any future handling a most unpleasant prospect.

I confess to a sort of perverse pleasure in watching my field crew—generally tough, young fellows—as their faces twist in revulsion at the lubbers' defensive tactics. These experienced outdoorsmen who routinely field dress deer handle the grasshoppers gingerly. They are, along with me, disgusted. What is this emotion and how can something the size of a cigar butt so powerfully infest our minds?

WHAT IS DISGUST?

Disgust is a universal human emotion that functions to protect the physical and psychological "self." We are disgusted by stimuli associated with contamination or infection. Our bodies can be invaded by all sorts of materials, chemicals, and organisms, so reacting negatively keeps us from contacting hazards and allows us to expel offensive material if it gets past our defenses. With cognitive sophistication, even ideas can be disgusting. For example, we find bestiality to be repulsive, as if a "dirty thought" could contaminate our consciousness. As Jimmy Carter famously confessed, "I committed adultery in my heart many times."[1]

The etymological origin of the word *disgust* points us toward the sense of taste. The word comes from *des*, meaning "the opposite of," and either the French *gout* or the Latin *gustare*, meaning taste (as in *gusto* and *gustatory*). This would be the end of the story, except in other languages the terms that are translated as *disgust* lack this connection to taste. The German *widerlich* (disgusting) and *Ekel* (disgust) connote a sense of being in opposition to something, along the lines of the Spanish *repugnante*, which is rooted in the Latin *pugnare*, meaning to fight (as in *pugilism*). In these cases, there is no implicit sensory quality. As such, some scholars believe that we might overemphasize the role of (dis)taste in understanding the nature of disgust.[2] Rather, the focus should be on the capacity of something to evoke a strong aversion due to its potential to taint through proximity, contact, or ingestion (only the last of these being explicitly a matter of tasting). Whatever the linguistic story, it is clear that English has an embarrassment of riches when it comes to words that convey a sense of disgust: *abhorrent, execrable, foul, gross, gruesome, nauseating, odious, offensive, putrid, repugnant, repulsive, revolting, sickening,* and *vile*—and all of these have been used to describe insects.

Although disgust got its start in the English language in the 1600s, scientific interest in this emotion didn't arise for another two centuries, with none other than Charles Darwin. While on his famed voyage, he observed:

> It is curious how readily this feeling is excited by anything unusual in the appearance, odour, or nature of our food. In Tierra del Fuego a native touched with his finger some cold preserved meat which I was eating at our bivouac, and plainly showed utter disgust at its softness; whilst I felt utter disgust at my food being touched by a naked savage.[3]

Philosophers and psychologists latched onto disgust in the early twentieth century, although interest largely faded as the study of other emotions became of greater interest and respectability. The exception was the psychoanalysts, who remained fascinated and interpreted disgust as a means of inhibiting the consummation of repressed urges.[4] In the past few years, however, disgust has enjoyed a scientific renaissance and seems to be emerging as the "white rat" of emotions, given how convenient and easy it is to elicit (e.g., a cockroach floating in a cup of tea)—and observe.[5]

* * *

While it takes considerable training to detect whether a person is lying, we have a rather easy task when it comes to knowing when another person is disgusted.[6] Shuddering and turning away with squinted eyes, wrinkled nose, and pursed lips are sure signs. If the object of disgust remains nearby, the individual may begin to gag or vomit. After establishing a safe distance, a disgusted person typically seeks to remove, cover, or clean up the stimulus. All of these reactions evidently function to keep the odious material from being near, getting into, or remaining inside one's body. And what's going on inside helps explain our overt responses.

Neurophysiology reveals that the experience of disgust is surprisingly distinct from that of its aversive cousin, fear. When we are disgusted, the parasympathetic nervous system is activated and our heart rate drops, whereas fear involves the sympathetic nervous system and elevates the pulse. Studies of brain activity in people watching film clips of hundreds of crawling cockroaches or images of people eating grasshoppers reveal that disgust is associated primarily with the anterior insula rather than the amygdala, which is activated during fear.[7] Interestingly, parts of the insula are within the gustatory cortex, where neurons associated with taste reside—so perhaps etymology reflects anatomy. However, the notion of the insula as a specific disgust processor is undermined by magnetic resonance images revealing that this region of the brain is also active during fear. Sorting out which part of the brain does what is complicated by the fact that a teeming mass of insects evokes multiple responses. There is, however, one feature of disgust that is unique.

TRIGGERING DISGUST: SENSATION AND COGITATION

No other emotion is so intimately linked to sensory experience as is disgust. We can recall and imagine instances of joy, sadness, surprise, fear, and anger—and in so doing, we may reexperience these occasions. Not so with disgust. As the scholar Robert Rawdon Wilson put it, "The representation of filth is not filth."[8] In other words, only an *actual* lubber grasshopper is disgusting. While René Descartes grounded our existence in the intellect through his famous dictum *Cogito ergo sum* (I think, therefore I am), we might claim to be certain that other things exist by virtue of *Fastidium movet, ergo est* (It is disgusting, therefore it is real). But the visceral reality of disgust is not drawn evenly from the five senses.

Given that disgust functions to protect us from contamination, it makes sense that smell drives our emotional response.[9] Odors warn us of rot, pollution, and disease before we come into physical contact with their sources. We smell a carcass long before we see it—and various creatures from stink bugs to skunks have exploited our capacity to be repulsed at a distance. Even without physical contact, inhaling vaporous emissions gives us a sense of having incorporated something of the vile object, and this intimacy can be nauseating. In earlier times, humans believed that foul odors were themselves carriers of disease. People sought to ward off the plague with a "pocketful of posies," which might justify a chuckle of modern superiority except that the makers of today's cleaning products exploit the same connection. The deodorant industry exploits a different sensibility. Because olfaction is perceived as a primitive, animalistic ability (being how mosquitoes find us and moths find mates, for example), today's humans associate smelling good with having no odor at all.

Smell is related to taste in that both are forms of chemical detection. And much of what we describe as taste is informed by smell—hence the grade-school experiment of tightly sealing a kid's nose and feeding him a piece of sweet onion and then a chunk of apple (similar textures and sugar contents) to show that taste alone cannot discriminate between these. Interestingly, the most aversive taste, bitterness, is not disgusting in itself. Instead, the only taste capable of eliciting disgust is sweetness in cloying excess.[10] Drinking tea with a drop of honey is appealing, but drinking a cup of honey is nauseating. The lesser role of taste in disgust makes sense (etymology notwithstanding) in that once something is in your mouth, contamination is all but assured. Of course, you can spit out a nasty caterpillar. Failing that, vomiting is a last resort after the viscous thing has coated your mouth in mucous slime and slid

Figure 4.2
Detail of the inner right wing of *The Garden of Earthly Delights* triptych by Hieronymus Bosch, painted in the 1480s. This panel depicts hell using grotesque and debased figures. The center character is a bird-headed monster sitting on a chamber pot and eating corpses, which are then excreted, while naked humans add their vomit and feces to the vile pit (image by Wikimedia Commons).

down your throat. However, these sensations are neither smell nor taste but the other highly relevant trigger of disgust.

We are keenly attuned to the tactile properties of substances that are likely to infect us—curdled, gooey, lukewarm, moist, mucky, oily, scabby, slimy, slithery, and squishy.[11] These are the textures of feces, mucus, lesions, innards, worms, snakes, cockroaches, and maggots. When kids want to evoke disgust on the playground, they often resort to a great American folk song handed down through the generations:

> Great, green gobs of greasy, grimy, gopher guts,
> Mutilated monkey meat, itty bitty birdie feet.
> Great, green gobs of greasy, grimy, gopher guts,
> And me without a spoon.[12]

When psychologists want to elicit disgust, they avail themselves of invertebrates. Taking a page out of the television show *Fear Factor*, various experimental protocols have subjects immerse their hands in a bowl of earthworms or hold a dead American cockroach.[13] Although touching is not as risky as tasting, it is surely more intimate than the remaining two senses.

Disgust at the sight of gray-green-purple rotting flesh or the sound of retching does not result from these sensory experiences per se, but from prior associations. Sight and sound reveal the existence of a vile object at some distance but do not "present" it with the intimacy of contamination. While disgust through the other senses has a noncognitive immediacy (e.g., we don't need to figure out what is causing a vile odor to be offended), sight and sound involve thought.[14] We do not find the colors and shapes in an unflushed toilet or the rustling of cockroaches in the dark displeasing without background knowledge and experience. An exception could be the auditory "impressions made by the pullulation of swarms of creeping insects,"[15] but even here we might suspect a cognitive element.

In sum, the consensus among psychologists is that the senses eliciting disgust are those that entail proximity to a source of filth. As such, smell, taste, and touch are the primary triggers of disgust, with vision and hearing eliciting this emotion by virtue of our memory or imagination. But humans are highly cognitive creatures, so to understand the ways in which insects induce disgust we must consider how thought and feeling conspire.

* * *

Although rooted in brute sensation, disgust is a "cognitively sophisticated emotion" that draws on our well-developed concepts of contamination and contagion.[16] We are disgusted by what we think something is and where we believe it has been. Such perceptions explain why people are reluctant to eat imitation dog feces fashioned out of chocolate or to drink a beverage that has been stirred with a comb.[17] Disgust arises from two strange but compelling psychological principles.[18] The law of contagion holds that "once in contact, always in contact." In this form of sympathetic magic, if something touches a disgusting thing, then it becomes disgusting itself (perhaps through what has been called "intuitive microbiology").[19] The law of similarity entails that benign things that look like foul things are rendered nasty. Applying these principles to insects, some larvae look like bird droppings (law of similarity), and we know where that fly on our hamburger has been (law of contagion).

The cognitive equivalent of too much sweetness can elicit disgust. For example, when there is an excess of flattery, we find such fawning to be objectionable and may describe it as "ass-kissing" or "brown-nosing" to express our revulsion. Likewise, those who muse over a world dripping with spiritual depth are said to be saccharine, while those who engage in unrestrained intellectualizing can become tedious—and monotony itself can evoke a kind of disgust.[20] These mental states are united by a kind of surfeit, which is most powerfully exemplified by an excess of sexuality or vitality.[21] There is something oddly similar in my not wanting to hear the details of your wedding night and your not wanting to hear my description of thousands of Mormon crickets seething onto a dirt road to feed on the greasy entrails of their brethren that were in the path of a vehicle. Too much of a good thing—or perhaps most any entomological thing—is offensive.

In fact, you probably expect my story of the teeming masses of crickets to be disgusting—a phenomenon called "interpretation bias." In experimental studies, subjects were shown pictures and asked to predict which of three outcomes would follow each image (tasting a disgusting liquid, feeling a shock, or nothing). Individuals who described themselves as extremely arachnophobic anticipated receiving either the shock or a nasty drink, the latter of which suggests that disgust might play a role in aversion to spiders.[22]

Pushing the cognitive aspect of disgust further, scientists have discovered a tendency of people to focus on offensive objects. The phenomenon of "attentional bias" is measured using a clever protocol known as the Stroop task. A subject is asked to tell the researcher the color of a word's text (e.g., an individual seeing the word *maggot* in blue letters says "blue"). When words associated with disgust are used in this task, people take longer to report the color than they do with neutral words, indicating the difficulty of ignoring a word's meaning. And when subjects are given a "disgust primer," such as hearing about a cockroach crawling into someone's mouth, the latency is markedly extended.[23] Such experiments along with case studies reveal the bizarre capacity of disgust to both draw our attention toward and push our thinking away from a stimulus.

In his darkly fascinating book on monsters, Stephen Asma recounts a conversation between a mother and her son at a medical museum. The boy was entranced by a display of a human fetus with two heads. When the mother asked, "Is this disturbing to you, William?" he replied, "God, yes. Very." But when she suggested they leave, the boy replied, "No, absolutely not."[24] I know how he felt. When I returned to the grasshopper-filled draw, I was torn between wanting to descend into the appalling superabundance of crapping, crawling copulation and wanting to run from the grotesque gulch.

We have all felt this Janus-faced phenomenon of allure and repulsion. Who hasn't looked while passing the aftermath of an auto accident? Indeed, the entertainment industry banks on our perverse curiosity. What is a horror film if not a manifestation of our prurient capacity to savor the abhorrent? But such a conflicted response does not require sensations as dramatic as those stimulated by watching zombies eat brains.

Recall that disgust plays out most directly through taste, smell, and touch. In earlier times, gourmets used the process of decay as a culinary tactic. The *haut goût* or "high flavor" lurking at the edge of revulsion was said to enhance the fleshy taste of meat.[25] Although few people today choose to eat putrefying flesh, many of us relish cheeses that are tinged (or even saturated) with the nidorous smell, pungent taste, and soft texture of spoilage—Stilton, Limburger, Pont l'Évêque, and Époisses (which even the French banned from being taken on public transport).

William Miller, a scholar of disgust, proposes two forms of this emotion—both of which shed light on our internal conflict. Freudian disgust combines with shame to serve "as a barrier to satisfying unconscious desires, barely admitted fascinations, or furtive curiosities."[26] Hence, we emotionally abhor what we secretly desire. And the disgust of surfeit protects us from overindulgence. Too much food, drink, sex, or other carnalities evokes nausea, so that which was attractive becomes repulsive. Perhaps Miller's model explains, at least partially, the push and pull of pornography—as well as of maggot-filled corpses, carpets of cockroaches, and legions of locusts.

THE DOMAINS OF DISGUST: INSECTS RULE!

Psychologists, philosophers, and other scholars have classified disgust in various ways. To understand why insects are so damn good at being disgusting, a biologically based taxonomy is most appropriate. Paul Rozin, the leading experimental psychologist in disgust who laid the foundations for this field in the 1980s, identifies seven "species."[27]

Animalism: Insects as Beastly Vectors

Rozin contends that disgust can be a manifestation of a desire to avoid our bestial origin and nature.[28] And according to Aurel Kolnai, a Hungarian philosopher whose thoughts on disgust in 1929 anticipated much of today's work, insects evoke a "strange coldness, the restless, nervous, squirming, twitching vitality [that gives] the impression of life caught up in a senseless, formless surging."[29]

Graham Davey, a psychologist, formulates animalism in terms of infection.[30] This connection between animals and disease arose in three ways. First, disgusting animals may possess the qualities of contaminating substances such as feces and mucus (e.g., worms and slugs are slimy and turd-like). Next, humans have correctly associated certain animals with illness (e.g., rats and cockroaches in our homes) and contamination (e.g., beetles in our grain and maggots in our meat). And finally, we have also falsely associated animals with sickness, as superstitions transformed some creatures into objects of disgust. For example, during the Middle Ages spiders were thought to absorb poisons from the environment and infect foods by contact (the difference between a chemical poison and a biological pathogen was not understood at the time). Spiders were also considered harbingers of the plagues that devastated Europe, with people believing that spiders (rather than fleas) spread disease through their bites. Indeed, some historians trace the emergence of our anxieties about spiders to the mistaken beliefs of medieval Europeans. This connection of animality to disease brings us to the second species of disgust.

Death: Insect Ambulance Chasers

The disgust evoked by teeming masses of insects arises from two morbid associations. First, rotting tissues fuel an outpouring of insect life, as if putrescence is reanimated in the form of flies, beetles, and their kin.[31] Indeed, until Francesco Redi's experiments in the 1600s, rotting meat was thought to generate flies spontaneously (and garbage was taken to be the source of rats). We have given up these beliefs, but it is still amazing to see blow flies arrive within minutes of death, as if these two-winged vultures are always lurking in the crevices of the world.

The other connection to death develops ironically from the profligacy of insects. In "senseless, formless surging" numbers, grasshoppers within a ravine represent the utter devaluation of life. Forced to confront our own triviality, we are appalled by the meaningless life and thoughtless death within the swarm. And so Miller contends that: "What disgusts, startlingly, is the capacity for life. . . . Images of decay imperceptibly slide into images of fertility."[32] From here, it is a small step to the next species of disgust.

Sex: Insects as Fornicators and Exhibitionists

An abundance of life implies a corresponding profusion of copulation. The orgiastic reproduction of insects has long offended human sensibilities: "Every swarming thing that swarms upon the earth is an abomination" (Leviticus

11:41). Kolnai described the disgust we feel toward vermin in terms of the "formless effervescence of life, of interminable directionless sprouting and breeding."[33] He maintained that repugnance is elicited "by the sight of swelling breasts, by swarming broods of some species of animal, fish-spawn, perhaps even by rank, overgrown vegetation."[34] Otto Weininger, another psychologist-philosopher in the early twentieth century, starkly claimed: "All fecundity is simply disgusting."[35]

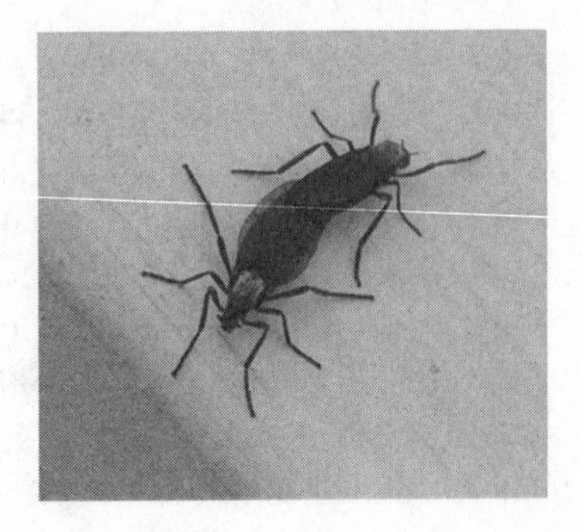

Figure 4.3
The March fly, *Plecia nearctica*, is commonly seen in enormous numbers along the Gulf Coast in spring and fall. The common name of love bug refers to the fact that these insects are often seen *in copula*, even while flying, thereby creating the impression of an insectan orgy. Such licentiousness in the animal world evokes disgust with exhibitionism, sexuality, and fecundity (image by Wikifrosch through Creative Commons).

In an affront to puritanical sensibilities, dragonflies, grasshoppers, and butterflies are seen *in copula* throughout the summer. In spring and fall across the southern United States, March flies (aka love bugs, honeymoon flies, and double-headed bugs) unashamedly mate on the wing. Appropriately enough, these six-legged exhibitionists spend their larval lives in the soil, consuming decaying vegetation. Indeed, we conceptually equate sexual license with dirtiness (e.g., pornography is "filth"), and this leads us to the next form of disgust.

Hygiene: Insects as the Original Dumpster Divers

If insects live in, consume, and emerge from sewage and garbage, it is easy to understand our revulsion. Hugh Raffles's lyric essay includes "the nightmare of long, probing antennae from the overflow hole in the bathroom sink or, worse, the rim of the toilet."[36] The nasty cockroach emerging from the plumbing is arguably more disgusting, but rather less common, than a germy fly crawling on our potato salad.

Our revulsion toward flies was intensified by Public Health Service programs in the early 1900s that rechristened the house fly as the "filth fly."[37] At about that time, Mark Twain was writing about the fly that coats itself with germs upon wading in festering sores and "then comes to the healthy man's table and wipes these things off on the butter and discharges a bowel-load of typhoid germs and excrement on his batter cakes."[38] Add to this the common knowledge (vividly captured by filmmakers in the remake of *The Fly*) that these insects regurgitate onto solid food to initiate the digestive process, and it's easy to understand why a fly on our meal is so gross. This takes us to our next species of disgust.

Food: Insects as Inedible Contaminants

Although insects are important foods in many societies, eating insects violates the sensibilities of the Western palate (even entomophagous cultures are rather discriminating as to which insects are on the menu). So offensive are insects in terms of the American diet that even traces of their bodies are scrupulously regulated. The US Food and Drug Administration considers insect parts on a par with rat droppings. For example, standards for a fifty-gram aliquot of cornmeal limit the number of insects to one, the amount of "insect filth" to fifty fragments, and the quantity of rodent filth to two hairs or one "excreta fragment."[39] So it appears that a grain beetle is comparable to a rat turd in your muffin—an equivalence that surely reflects the emotion of disgust more than the rationality of science. Regulations aside, we might prefer a couple of rodent hairs over the excrement—except for our next form of disgust.

Bodily Products: Insects as Yucky Stuff

Hair, feces, urine, mucus, saliva, sweat, blood, vomit: this is the stuff of primal, visceral disgust. At least most insects aren't hairy (furry caterpillars notwithstanding), so aside from defecating, regurgitating lubber grasshoppers, insects don't generally contaminate our world with their bodily products. However, the final species of disgust is another matter.

Bodily Violations: Insects as Invaders

Last summer, I was sitting with my son while he was being prepped for surgery to reassemble his shattered collarbone—a rather grisly bodily violation, in my queasy estimation. Out in the hall, I heard a brief ruckus and my son's nurse say, "Oh thank you! I couldn't do that. It just turns my stomach." I peeked around the corner and saw that a medical technician had squashed a cricket.

Insects in hospitals, metal screws in bones—transgressions of boundaries. Clinical psychologist Susan Miller argues that the greater the potential for something to enter us, the greater the disgust:

> Small, primitive life-forms close at hand are especially likely to disgust us. I believe this is because they seem too likely to enter us or at least to latch on. . . . they seem hungry for an affiliation with something more substantial. If they are structurally designed to cling or ooze, the problem worsens.[40]

Lice infesting pubic hair and worms squirming from an anus are paradigm cases of creatures violating our boundaries—of insinuating, transgressing, trespassing. Our essential "self" is compromised when our biological or psychic skin is breached. We might also feel revolted when we are the violators, as when we penetrate an amorphous, protean mass of grasshoppers.

* * *

Having considered these species of disgust, we might wonder whether insects would fare as well (or badly, depending on one's perspective) with other taxonomies. Does Rozin have it in for these creatures—or is there something about them that any cataloging would reveal? Let's conclude with a whirlwind tour through the paired terms that William Miller uses to classify disgust (which he associates with the second descriptor in each pair): inorganic/organic, plant/animal, human/animal, us/them, me/you, outside of me/inside of me, dry/wet, fluid/viscid, firm/squishy, nonadhering/sticky, still/wiggly, uncurdled/curdled, life/death-decay, health/disease, beauty/ugliness, up/down, right/left, cold-hot/clammy-lukewarm, tight/loose, moderation/surfeit, one/many.[41] We might quibble about just how many of the latter terms pertain to insects, but it seems reasonable to describe many of "them" as organic, squishy, sticky, wiggly, ugly, animals associated with surfeit, death, decay, and disease.

So we see that disgust arises from a complicated set of sensory experiences and cognitive associations. However, not only the triggers of disgust but the feeling itself is, well, sloppy. Like a sticky, mucous substance, disgust is difficult to separate from other emotions. However, coming to understand the entanglements is vital to understanding the infested mind.

NOTES

1. Amanda C. Weldy, "A thoughtful call for higher moral standards, courtesy of *Playboy* magazine," *Journal of College and Character* 10, no. 4 (2009): 1–3.
2. Susan B. Miller, *Disgust: The Gatekeeper Emotion* (Hillsdale, NJ: Analytic Press, 2004).
3. Quoted in William Ian Miller, *The Anatomy of Disgust* (Cambridge, MA: Harvard University Press, 1997), 1.
4. Aurel Kolnai, *On Disgust*, ed. Barry Smith and Carolyn Korsmeyer (Chicago: Open Court, 2004), 3.
5. Miller, *The Anatomy of Disgust*; Miller, *Disgust: The Gatekeeper Emotion*; Bunmi O. Olatunji and Dean McKay, eds., *Disgust and Its Disorders* (Washington, DC: American Psychological Association, 2009); Daniel Kelly, *Yuck! The Nature and Moral Significance of Disgust* (Cambridge, MA: Bradford/MIT, 2011); Carolyn Korsmeyer, *Savoring Disgust: The Foul and the Fair in Aesthetics* (New York: Oxford University Press, 2011); Colin McGinn, *The Meaning of Disgust* (New York: Oxford University Press, 2011); Rachel Herz, *That's Disgusting: Unraveling the Mysteries of Repulsion* (New York: W. W. Norton, 2012).
6. Kolnai, *On Disgust*, 33–35.

7. Anne Schienle, "The functional neuroanatomy of disgust," in Olatunji and McKay, *Disgust and Its Disorders*, 147–54.
8. Quoted in Nat Hardy, "Imagining disgust," *Canadian Review of Comparative Literature* 34 (2007): 424.
9. Miller, *The Anatomy of Disgust*, 66–77.
10. Ibid., 85–87.
11. Ibid., 38.
12. "A Fish That's a Song," track 11 sung by Mika Seeger on Smithsonian Folkways Recordings, 1990.
13. Bunmi O. Olatunji and Josh M. Cisler, "Disgust sensitivity: Psychometric overview and operational definition," in Olatunji and McKay, *Disgust and Its Disorders*, 45–46.
14. Kolnai, *On Disgust*, 48–52.
15. Ibid., 52.
16. Miller, *The Anatomy of Disgust*, 6.
17. Ibid.
18. Paul Rozin, Jonathan Haidt, and Clark McCauley, "Disgust: The body and soul emotion in the 21st century," in Olatunji and McKay, *Disgust and Its Disorders*, 14–15; Nathan L. Williams, Kevin M. Connolly, Josh M. Cisler, Lisa S. Elwood, Jeffrey L. Willems, and Jeffrey M. Lohr, "Disgust: A cognitive approach," in Olatunji and McKay, *Disgust and Its Disorders*, 59.
19. Steven Pinker, *How the Mind Works* (Harmondsworth: Penguin, 1998).
20. Kolnai, *On Disgust*, 62–72.
21. Ibid.; Miller, *The Anatomy of Disgust*, 40–47, 109–12.
22. Peter J. de Jong and Madelon L. Peters, "Contamination vs. harm-relevant outcome expectancies and covariation bias in spider phobia," *Behaviour Research and Therapy* 45 (2007): 1271–84.
23. Williams et al., "Disgust," 61.
24. Stephen T. Asma, *On Monsters: An Unnatural History of Our Worst Fears* (New York: Oxford University Press, 2009), 6.
25. Kolnai, *On Disgust*, 16–22.
26. Miller, *The Anatomy of Disgust*, 109.
27. Paul Rozin, Linda Millman, and Carol Nemeroff, "Operation of the laws of sympathetic magic in disgust and other domains," *Journal of Personality and Social Psychology* 50 (1986): 703–12; Paul Rozin and Carol Nemeroff, "The laws of sympathetic magic: A psychological analysis of similarity and contagion," in *Cultural Psychology: Essays on Comparative Human Development*, ed. James W. Stigler, Richard A. Shweder, and Gilbert Herdt (Cambridge: Cambridge University Press, 1990).
28. Miller, *The Anatomy of Disgust*, 6.
29. Kolnai, *On Disgust*, 58.
30. Graham C. L. Davey and Sarah J. Marzillier, "Disgust and animal phobias," in Olatunji and McKay, *Disgust and Its Disorders*; Graham C. L. Davey, "Characteristics of individuals with fear of spiders," *Anxiety Research* 4 (1992): 299–314.
31. Kolnai, *On Disgust*, 16–22; Miller, *The Anatomy of Disgust*, 40–41.
32. Miller, *The Anatomy of Disgust*, 40.
33. Kolnai, *On Disgust*, 62.
34. Ibid.
35. Quoted in ibid.
36. Hugh Raffles, *Insectopedia* (New York: Pantheon, 2010), 202–3.
37. Charlotte Sleigh, "Inside out: The unsettling nature of insects," in *Insect Poetics*, ed. Eric C. Brown (Minneapolis: University of Minnesota Press, 2006).

38. Mark Twain, *Letters from the Earth*, ed. Bernard DeVoto (Greenwich, CT: Fawcett Crest, 1964), 35.
39. US Food and Drug Administration, "Defect levels handbook: The food defect action levels," http://www.fda.gov/Food/GuidanceRegulation/GuidanceDocuments RegulatoryInformation/SanitationTransportation/ucm056174.htm (accessed March 29, 2012).
40. Miller, *Disgust: The Gatekeeper Emotion*, 56.
41. Miller, *The Anatomy of Disgust*, 38.

CHAPTER 5

The Maggoty Mind: A Natural History of Disgust

An entomologist presented with a new specimen faces a challenge not unlike that faced by a psychologist in a first session with a patient who has an emotional disorder. Three questions tend to focus the professional's attention.

First, a preliminary identification is in order. Given that there are few million species of insects, the entomologist has to narrow the field. Is the creature a blow fly maggot, a carrion beetle larva, or a clothes moth caterpillar? Likewise, sorting out a patient's emotions can be a challenge—is the individual experiencing disgust, fear, or contempt?

Next, to solve a problem, the entomologist often finds it necessary to discern the origin of the infestation. Knowing whether the specimen came from a person's basement, bedroom, or body matters. And for the psychologist, how a patient's revulsion arose can be a valuable part of deciding on a course of treatment.

Finally, the entomologist has to deal with sometimes bizarre variations. I remember my first encounter with gynandromorphy. Identifying grasshopper species (a key to sound pest management) often requires a careful examination of diagnostic features of the male genitalia—and it's quite confusing when the right half of a specimen is male and the left side is female! I can imagine the surprise of the psychologist whose patient is disgusted by insects but who has transferred this emotion into a form of sexual stimulation.

These are our three challenges in understanding the nature of disgust. And it turns out to be a very messy undertaking.

THE EMOTIONAL FAMILY TREE OF DISGUST

Disgust's closest relative is surely fear, as these are the two aversive emotions that protect us from harm. Much of what we know about the workings of fear comes from the study of individuals who respond negatively to insects and spiders. However, it is becoming less clear that such individuals are simply or solely afraid.[1] Perhaps, rather than thinking that people have entomophobia or arachnophobia, we should recognize their reactions as a complex aversion best called an *apostrophe* (Greek for "turning from"). So entomapostrophe might be an accurate, if inelegant, term for a response to insects based on fear, disgust, or both.

Among arachnophobes (sticking with the traditional term), there is a consistently strong link between disgust sensitivity and response to spiders. Following successful treatment, not only are patients less fearful but they also report a marked reduction in disgust toward spiders.[2] While arachnophobia is evidently a mixture of fear and disgust, what about more typical responses to multilegged creatures? Surely normal folks can parse their emotions, right? Wrong—even people without debilitating reactions to insects and their relatives tend to exhibit muddled responses to these animals. In the general population, individuals' views about the ability of spiders to spread contamination (typically a feature of disgust) are related to the extent of their fear of spiders.[3]

Given the snarled relationship between disgust and fear, scientists have sought to explain why and how these two experientially distinct emotions are so difficult to isolate in practice. Nobody would feel disgust upon coming to the edge of a precipice or be afraid upon finding a pool of vomit. But when it comes to a cockroach under the sink or a spider in the basement, we seem to be flummoxed. Five frameworks have been used to explain our confusion.

First, the "imprecision model" is based on psychologists' having found that people are often unsure about how to label their emotions.[4] Although women differentiate fear and disgust more carefully than do men, both sexes tend toward vagueness when emotions are moderate. Perhaps some experiences are unambiguous by virtue of their intensity (e.g., arriving at a precipice). In other cases, what appear to be clear differences in principle (disgust and fear) are muddled in practice (e.g., the pool of vomit may immediately evoke disgust, with fear of contracting an illness developing as we think about the situation). And insects often present complex stimuli giving rise to complicated emotions.

Another explanation is the "synergy model."[5] Even when fear and disgust are distinguishable, these emotions may feed on each other. Researchers have found that fear of spiders amplifies disgust; subjects report greater disgust when fear is concomitant. And the reverse can also occur. Psychologists posit that individuals who are disgusted by insects understandably avoid proximity to these organisms; hence they also miss opportunities for harmless experiences to undermine their fear, which may develop unchecked into a full-blown phobia.

Third is the "fear first model," which holds that disgust originates in fear.[6] According to this theory, fear is a response to either danger or contamination—and fear of the latter is disgust. This view also provides insight into one of the most common manifestations of obsessive-compulsive disorder.[7] Repeated cleaning is a response to the fear of contamination, and such ritualized cleansing tracks gender differences in disgust sensitivity. Moreover, exhibiting a fearful response to disgusting animals is positively correlated with obsessive-compulsive washing.

The next model inverts the previous psychological formula. The frequency with which high levels of disgust have been found in people suffering from small animal phobias in general and arachnophobia in particular has led some psychologists to propose an "aversion genesis model" in which fears are rooted in disgust.[8] The idea is that if an individual is experiencing disgust, then she is likely to interpret movements and other behaviors on the part of the disgust-inducing creature as threatening, thereby setting the stage for fear. The preponderance of the scientific evidence supports such an emotional mutualism.

Compared to their nonphobic peers, arachnophobic children are both more prone to disgust in general and significantly more disgusted by spiders. Interestingly, the best predictors of a highly aversive response to spiders by children are the creatures' disgust-evoking qualities, rather than their threatening features.[9] Salvador Dalí traced one thread of his entomophobia to a childhood experience of disgust. Upon catching a slimy fish, he cried out in terror:

> My father, who was sitting on a rock nearby, came and consoled me, trying to understand what had upset me so. "I have just looked at the face of the 'slobberer,'" I told him, in a voice broken by sobs, "and it was exactly the same as a grasshopper's!" Since I found this association between the two faces, the fish's and the grasshopper's, the latter became a thing of horror to me.[10]

Finally, the "horror model" takes us back to the beginning of our journey into the infested mind. This framework—exemplified by Dalí—recognizes that sometimes neither "fear" nor "disgust" is a sufficient label for our emotional response.[11] For example, upon encountering a greasy cockroach, we may perceive that there is no escape physically or psychologically. If we take flight (fear typically entails removing ourselves from the object) and leave the room in fear, the creature will remain in our kitchen. And if we try to step on the insect (disgust typically entails removing the object from our presence), the creature is likely to evade us and squirm beneath the refrigerator. Even if we succeed in crushing it, with the attendant slippery gooiness and stale-urine stench, we know that there is never a single cockroach. In any case, its being out of sight does not make it out of mind. No matter what we do, its filthiness is inescapable. The term for such fear-imbued disgust is *horror*—the sense of being repulsed without being able to distance ourselves in body or mind. In Sue Hubbell's eminently readable *Broadsides from the Other Orders: A*

Book of Bugs, we find a vivid account of horror in a college professor's description of his response to cockroaches:

> I see one in the kitchen and I am terrified, paralyzed, unable to speak or move. It is so small and I am so big. That is part of the horror. It is not the least afraid of me. . . . Kill one? That would be impossible. It is, psychically, too big to kill. . . . Besides, even if I were able to kill it, another would come, another just like it. And that is too frightening even to consider.[12]

* * *

There is no doubt that disgust and fear are often entangled, but psychologists have developed some clever methods for teasing apart these emotions. Imagine going into the kitchen for a snack of tea and cookies. After making a cup of Earl Grey, you turn and see a spider crawling across the snickerdoodles. Sensing your movement, the spider dashes off the table and disappears into a heating vent. Now what? Do you shrug and eat a cookie? Do you give the cookies a perfunctory wipe and munch away? Or do you refuse to eat the cookies—maybe even throw them out?

The spider-walking-on-the-cookies test is a simple approach to figuring out whether someone is merely afraid of spiders.[13] If so, he or she will eat the cookies once the spider is gone. No spider, no fear, no problem. However, for someone who is disgusted by spiders, even once the creature has left, its contaminating qualities linger—and the cookies remain repulsive. But we're not done yet. What about the cup of tea?

When psychologists served phobic (spider-disgusted) and nonphobic subjects tea in a grungy cup, arachnophobes were no more likely to reject the beverage.[14] The use of a dirty cup served as an indicator of disgust sensitivity, and it appears that the revulsion that these people felt toward spiders did not generalize. So with regard to insects and their kin, not only do disgust and fear come apart for some people, but also the sense of heightened disgust is restricted to multilegged life forms.

Another way of teasing apart fear and disgust involves dissecting our responses to animals. Across cultures, aversions coalesce around two features: danger and contamination. The most frequently frightening brutes are tigers, alligators, lions, bears, sharks, and wolves, and the most commonly off-putting tramps are cockroaches, spiders, and slugs—along with vertebrates of their ilk such as snakes, bats, and rats.[15] Our responses are driven by our perceptions of the possibility of our being attacked or infected. Simply put, big creatures are likely to eat us and little creatures are likely to carry pathogens. In support of this interpretation, various studies have shown that we associate pain with "predators" (the violent group) and illness with "fear-relevant" animals (the nasty group).

So-called outcome beliefs (what we think will happen) appear to be key in differentiating fear of danger from fear of contamination. For any given creature, do people expect to be hurt or infected? Unfortunately, the answer is not so clear for some animals. In a carefully controlled experiment, spider-fearful individuals were asked to estimate the probability that an image of a spider, a maggot, or a pit bull would be followed by an electric shock (pain) or a sip of nauseating juice (disgust). Pit bulls were associated with a painful outcome and maggots with a nasty outcome. And the spiders? People anticipated both results, suggesting that the neat fear/disgust or danger/contamination dichotomy may not account for all of our emotional responses.[16]

* * *

Fear is not the only psychological relative of disgust. Psychologists have analyzed an extended family of emotions that make for a most unpleasant household—and a great challenge to diagnosis.[17] To begin, there is *contempt*, which is generally taken to be associated with pride, superiority, and indifference. The object of disgust is surely beneath us (maggots are not our equals), but our response is not really one of arrogant dismissal. Next we have *shame*, which can be understood as a kind of self-loathing. However, disgust is not normally directed toward one's own character or features.

Abhorrence is a close cousin of disgust. In this case, analysts differentiate these responses in terms of their immediacy and sensuality. When we encounter the cockroach there is no contemplative lag in terms of our disgust. However, with a bit of reflection as we search for the can of insecticide, we might well work ourselves into a state of abhorrence. That is, our knowledge of cockroaches can underwrite our negativity, but now we have moved from a primarily visceral to a largely cognitive state.

Moving on to greater intensity, we come to *hatred*. It is certainly possible for people to hate mosquitoes, for example. However, hatred is unrestricted in the sense that it can pertain to almost anything concrete or abstract, while disgust is associated with organic aspects of the world. Kolnai contends that our intentions are in the foreground of hate, while disgust is much more about the subject. Hatred is a sibling of *anger*, but this is also different from disgust, which is more primordial. Anger depends on our making judgments about something, while disgust is a matter of sensuously grasping qualities of an object or process.

The fraternal twin of disgust is *love*, which would appear to have nothing in common but is arguably the closest relative. William Miller contends that "love bears a complex and possibly necessary relation to disgust."[18] His contention is that while disgust functions to mark the boundaries of oneself, love holds the trump card. Simply put, sexual love allows us to make babies and nonsexual love allows us to change their diapers. And for entomologists, a kind of intimacy permits engaging insects with a kind of physical and psychological closeness that many people find hard to imagine—but perhaps this difficulty is no greater than that of a bachelor attempting to comprehend how a father can wipe his child's nose without disgust. For me, a deep familiarity with—even affection for—grasshoppers kept me engaged with these creatures after my horrifying encounter. Individually they were, and still are, enchanting to me.[19]

THE MAKING OF DISGUST: INSTINCTS, INSTRUCTIONS, AND INSECTS

We are not born disgusted. In fact, for the first two years of life, babies show no signs of disgust.[20] That is not to say that they don't exhibit distaste, which is considered a protoemotion in that it is a purely sensory response. And what babies put in their mouths is rather interesting. In what was presumably a challenging study to get through the human experimentation approval process, nearly two-thirds of diapered subjects ate imitation dog feces realistically crafted from peanut butter and stinky cheese, and nearly as many ate a small dried fish. However, only one-third ate a whole, sterilized grasshopper, and less than 10 percent munched on a lock of hair. Rejection of the latter substances may well have been a matter of "mouth feel" or texture as much as one of taste.[21]

A capacity for disgust develops between the ages of four and eight—a time in which a child is becoming more independent and less able to rely on parents to provide protection from harmful substances.[22] A study with children between eight and twelve years old showed increasing sensitivity to contamination with age. When researchers put objects of varying nastiness close to a tasty beverage and assessed children's willingness to take a drink, they found that candies, hot dogs, and leaves did not deter the kids, but the presence of a grasshopper, poison, or feces was off-putting.[23]

With sexual maturation, adolescents become exceptionally aware of and responsive to repulsive stimuli. As individuals age further, disgust sensitivity declines, along with the tendency for self-monitoring. Although this diminishment is greater in women, they continue to be more sensitive than men.[24]

The mechanisms through which we acquire disgust are similar to those for fear; we learn through direct experience, observational modeling, and communication of negative information. So it is not surprising that mothers of phobic girls are more likely than other parents to find spiders disgusting. However, yet another form of learning plays an important role.

Imagine that you're about to order breakfast at your favorite diner. After reading the menu, you settle on the pancakes but can't decide between orange and cranberry juice. While pondering this, you look across the counter and see a cockroach drop into a glass of orange juice. The waitress gives a shudder and pours the drink into the sink, and then she turns to you and asks for your order. Which drink do you prefer? You know that you'll get a new glass of juice, but it's likely that you'll go with the cranberry because you've now associated cockroaches with orange juice in a process called "evaluative conditioning."[25]

Just such a scenario has been played out experimentally, and people show a reduced preference for a fresh serving of an otherwise appealing drink if they've seen a cockroach in a glass of the same liquid.[26] What's more, when people are assured that the dunked cockroach is sterilized, they still reject the beverage. Our initial evaluation based on disgust is recalcitrant. Even a rational consideration of the situation does not typically allow us to rescind our judgment.

The implications of evaluative conditioning for treatment of entomophobia are rather troubling. Having patients handle maggots may ameliorate fear but it can also result in their attending to the insects' offensive qualities and a further elaboration—rather than diminishment—of disgust.[27]

* * *

If disgust is so readily and solidly entrenched in the human psyche, does this suggest that something more than individual experience is at play? Could our revulsion toward insects be rooted in the evolutionary recesses of the mind? The evidence that we are genetically predisposed to wrinkle our noses at maggots and cockroaches—along with grasshoppers that are puking, shitting, or swarming—is intriguing and perhaps compelling.

To begin, the facial contortions that we recognize as expressing disgust are common to most, and perhaps all, cultures.[28] Around the world, people recognize the "disgust face"; Mr. Yuk's green face is used to warn children away from poisons. Moreover, some stimuli elicit disgust across cultures. Feces do the trick, but insects also appear on everyone's list. Various studies have found that people from Burkina Faso, Hong Kong, India, Japan, Korea, the Netherlands, the United Kingdom, and the United States are all repulsed by the "standards" (e.g., flies, cockroaches, and spiders) along with some idiosyncratic responses (e.g., the Dutch are put off by aphids on lettuce). In terms of cultural sensitivities, Indians respond to fewer animals than do other nationalities, and the Japanese respond to more.[29]

Susan Miller contends that "disgust serves as the psychic equivalent of the cell wall."[30] This rapid aversive response is a protective mechanism that natural selection has shaped to inhibit our contact with substances that endanger us through infection or intoxication. From this biological perspective, we can see why the sense of smell is so powerfully linked to disgust, as it warns of proximity. And other behavioral phenomena also fall into place, such as developmental patterns.

Figure 5.1
Facial expressions and bodily postures associated with disgust from Charles Darwin's *The Expression of the Emotions in Man and Animals* (1872). Darwin argued that humans and non-human animals share certain psychological states, contrary to the earlier work of Charles Bell, who contended that divinely created muscles allowed the expression of uniquely human feelings (image by Wikimedia Commons).

As Miller puts it, "Humans obviously are engineered with a readiness to develop disgust as one of the affect-maturational trajectories of childhood."[31] Beginning with a lack of disgust in infancy makes sense in terms of evolution. When we are young, our parents protect us from foul and poisonous materials. Given the range of foods in the varied environments of humans, it is important that babies sample and come to accept those provided by their parents.[32] While there are striking variations in what cultures find delectable, there are also clear tendencies. For example, children can come to accept slimy foods, but substantial social investment is required to accomplish this.[33]

Evolutionary psychologists contend that prepared learning accounts for our tendency to readily associate certain animals with disgust.[34] Stinky, greasy, wriggling creatures have been particularly likely to transmit disease throughout human history, so we are primed to find cockroaches and maggots gross. Even less direct sources of revulsion may have adaptive advantages.

Surfeit, for example, may reflect on adaptation to the fact that too much of anything can put us at risk.

Although we think of disgust as primal, ethological studies have led scientists to believe that it may have been the last of the basic, pancultural emotions to emerge in the course of human evolution.[35] None of our primate relatives seem capable of disgust. Like human infants, they express distaste, but the full-blown emotion of disgust is not in their repertoire. Perhaps monkeys and their brethren lack the cognitive capacity for anticipating illness, but the difference may also lie in the uniquely human quality of self-awareness. If the truly defining feature of disgust is its protection of the self, then without a sense of identity or autonomy an animal might find an odor to be unpleasant but it cannot make the leap to existential violation. Perhaps William Miller is right when he proposes that: "To feel disgust is human and humanizing."[36]

But, of course, arguing is also human—and the cultural constructivists aren't about to cede disgust to the evolutionary biologists.

* * *

Anthropologists have documented a staggering array of animals, plants, and fungi that humans eat, and such diversity is taken as evidence that our disgust toward certain foods is a cultural artifact.[37] Why should eating a grasshopper be any more repulsive than drinking the milk of another species? More to the point, the roasted grasshopper is far less likely than the raw milk to pass an infection along to the consumer. As an aside, I have caught, cleaned, and cooked (baked, fried, and sautéed) grasshoppers and find their shrimplike flavor and texture to be quite pleasant—more so than the fatty viscosity of whole milk. However, I would submit that the sliminess of raw oysters is unlikely to be appealing to most any human without cultural approbation.

There is no doubt that we socialize children to what adults consider disgusting through both direct instruction and facial expressions. A mother makes a face and tells her child, "Yuck, put that down. You don't know where it's been" (meaning that it has almost surely been someplace dirty). However, training need not be explicit, and parents are often unaware that they are using the "disgust face" during interactions with their children.[38]

When a child latches onto a grasshopper, the modern parent's signaling of revulsion may reflect another cultural influence—a disconnection from the natural world to the extent that we are no longer able to distinguish one insect from another. Hence, all multilegged creatures are generalized into an undifferentiated lump of disgusting things.[39] With the advent of civilization, humans became able to discriminate among kitchen utensils, art genres, and religious symbols but lost the capacity to distinguish species. Today, more people can tell the difference between toy and real guns than between edible and poisonous spiders. So, the argument goes, our disgust is a product of enculturated ignorance—not biological evolution.

What about feces—isn't the universal disgust toward this substance evidence that evolution calls the shots? Not so fast. A study of feral humans revealed that without social interaction humans find fecal odors interesting and not repulsive.[40] That said, whether a society could adopt a positive affect toward feces, with children engaging this material as if it were Play-Doh, is quite another matter.

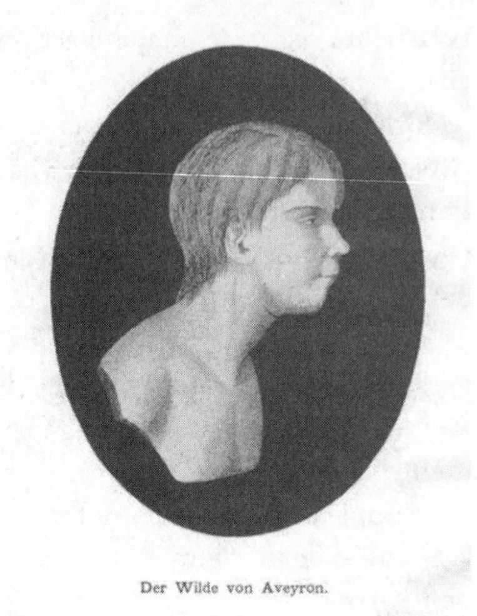

Figure 5.2
Victor, or "the Wild Boy," of Aveyron was a feral child in the late 1700s. At about age twelve, Victor entered human society and was studied by a physician who made important discoveries pertaining to the education of developmentally delayed children. Feral humans provide a window into the origins of emotions, and it appears that most objects of disgust involve some degree of enculturation (image by Wikimedia Commons).

What seems clear is that culture has exploited our capacity for disgust in the realm of ethical judgment. We speak of moral decay, filthy liars, revolting acts, foul characters, and lousy things to do (the last of these expressions is etymologically derived from the state of being infested with lice). Often our righteous revulsion is conceptualized and expressed in terms of insects. Indeed, we cast the net of moral condemnation over all of the invertebrates when we say that a person lacking integrity or courage is spineless. Human beings should rise above their vulgar, bodily desires, and when they fail to do so we equate them with animals—and in the Great Chain of Being insects are near the bottom. As Susan Miller puts it, "We often feel disgust toward animals when they are portrayed as creatures dominated by their drives, which are seen as base, unruly, and undignified."[41] When we perceive other people as vermin such as lice and cockroaches, contempt toward them is proper. As sources of contamination they are suitable targets for extermination, in the same sense that flies must be eliminated from a child's nursery.

Just how much latitude does a society have in terms of evoking disgust? Anthropological evidence suggests that we have considerably more leeway to include objects and actions than we have to exclude them.[42] But even inclusive freedom seems to have limits—no one considers pebbles, snowflakes, or thumbs disgusting. Even a few insects are difficult, if not impossible, to render disgusting (e.g., butterflies). To figure out a society's sensitivity to disgust, just count up the rules. Across cultures, the more norms a group has in a given realm, such as food or the body, the lower the threshold of disgust when a guideline is violated.[43]

In the name of tolerance, Western culture inculcates a kind of politically correct appreciation of nature in which many scholars now contend that all of nature is beautiful. Instead of judging organisms on our own terms, we are supposed to accept them for what they are. Rather than being perceived as the ugly old women of the prairie, lubbers are to be seen as elegantly colored, finely sculpted, noble creatures. However, when a humongous insect regurgitates and defecates (perhaps I should say "pukes and craps" to retain an earthy integrity) no amount of cultivated sensitivity saves this from being a disgusting encounter.

* * *

In the end, we are left with two stories about the making of disgust. The evolutionary psychologists tell us that our ancestors who found certain substances offensive avoided contamination, thereby producing offspring with the tendency to be similarly disgusted—along with the capacity to associate other things with the objects of disgust. Hence, we are genetically primed to be repulsed by feces and readily learn to connect flies with this nasty substance. The cultural constructivists tell us that our ancestors were like feral humans with no innate sense of disgust, but civilization brought high population densities and the need for hygiene, which we crafted into a new emotion. Hence, humans have passed down norms about feces and flies through socialization.

Daniel Kelly has proposed an integration of these views through a combination of his "entanglement thesis" (disgust being the emotional response to two functionally integrated mechanisms that protect us from ingesting foul substances and from becoming infected by pathogens or parasites) and the "co-opt thesis" (humans exploited the power of disgust to protect groups from violations of social norms and transgressions of social/spatial boundaries).[44] Perhaps William Miller resolved the debate most succinctly: "Silly things get said in favor of the extreme social constructionist position and even sillier things in favor of the social-biological extreme."[45] The point is that we are culturally malleable creatures operating within evolutionary constraints, and insects lie somewhere between puke and pedophilia.

And it is just this psychological flexibility that allows some extraordinary associations involving insects, disgust, and sex.

PARAPHILIA: THE PECULIAR PLEASURE OF ANTS IN YOUR PANTS

Paraphilia is the phenomenon in which a person is sexually aroused by objects or situations that are not deemed normal by society. Paraphiliacs depend on unusual stimuli that most often fall into one of two types: fear-based stimuli or disgust-based stimuli. The psychophysiology is rather straightforward. When something alters our heart rate and respiration, the mind interprets this change in the context of the situation.[46] If there is a lion or a cockroach before us, we infer fear or disgust, but if there is a gorgeous human next to us, we infer attraction. If you monitor the physiological state of a person without knowing the setting, it is difficult to discern whether that individual is engaged in bungee jumping or foreplay (hint: take your next date to a horror movie and then out for a double espresso).

So some people use scary and disgusting things to generate physiological arousal, which is then converted into sexual excitement. Paraphiliacs variously find heights (acrophilia), strangulation (asphyxiophilia), enclosed spaces (claustrophilia), theft (kleptophilia), public speaking (homilophilia), burial alive (taphephilia), and snakes (ophidiophilia) to meet their needs. As for disgust, people use feces (scatophilia), vomit (emetophilia), farts (flatuphilia), mucus (mucophilia), urine (urophilia), and animals (zoophilias include sexual encounters with dogs, horses, sheep, rodents, pigs, chickens, fish, and so on). And when we're talking about physiological arousal from fear, disgust, or both (remember, these emotions are often confounded), insects can be just the ticket.

* * *

Formicophilia is an uncommon but fascinating variety of zoophilia involving ants (the term being derived from the scientific name for ants: family Formicidae).[47] One might imagine that ants could evoke a sense of fear, given the pain associated with their bites and stings, or perhaps putting insects on one's body entails a sense of disgust, or maybe having little creatures clambering over one's erogenous zones just feels good. In any event, the classic case of this paraphilia reported in the psychological literature involved a twenty-eight-year-old Sri Lankan man.[48]

As a child, the fellow was something of a social misfit. He was a nerd—bright, shy, and averse to roughhousing. At the age of nine he began keeping ants, which in itself isn't unusual (I had several ant farms during my childhood, and I turned out pretty normal). However, he liked "the ticklish feeling [of having] the ants crawl on my legs and thighs."[49] OK, a little odd, but nothing too worrisome.

The turning point seems to have been a pair of traumatic events. First, a domestic servant of the family taught him how to masturbate, which wouldn't have been shocking except that his father caught them. The older boy was dragged down to the police station, throttled in front of the child, and then fired—after which the younger boy was severely beaten. Next came the death of the boy's mother from breast cancer and the domination of his life by the tyrannical father. By age eleven, the boy was increasingly isolated and depressed.

In his early teens, he began devoting more time to his "hobby" and expanded his collection to include snails and cockroaches. Now he was masturbating while the ants crawled on his legs, and the path to formicophilia was seemingly inevitable. He dropped out of school at eighteen and avoided women because he felt that they could discern "what a disgusting man I am."[50] And it didn't help that at this time he witnessed his father having sex with a family friend and was appalled that a woman behaved "just like an animal."[51]

For the next decade his paraphilia developed, until he was using insects for sexual arousal several times a week:

> After removing all his clothes he would take a few snails and cockroaches from his zoo cupboard. He would then lie on the floor and have the cockroaches crawl over his thighs and testicles, while the snails slid over his nipples and the tip of his penis. . . . He also enjoyed the sensation of the cockroaches crawling around his perineum and anus. On such occasions he would masturbate to orgasm as often as four or five times within the space of 2 or 3 hours.[52]

At age twenty-eight he sought psychological help for his paraphilia, weeping and begging to be saved from "my disgusting habit."[53] Fortunately, the psychiatric clinic at the University of Colombo provided compassionate treatment (historically, paraphiliacs have been castrated and subjected to aversive conditioning with strong electric shocks). Over the course of twelve weeks, the young man was taught basic social skills for interacting with women and encouraged to use heterosexual pictures for sexual arousal. By the end of the therapy, his interest in women had markedly increased and his reliance on insects had diminished.

Although the term *formicophilia* is etymologically linked to the use of ants, the definition has been expanded to "the use of insects for sexual purposes . . . to crawl over the genitals or other portion of the body."[54] And the choices of insects reflect the diversity of these creatures.[55] The earliest reported case was in the time before electricity—and hence before vibrators. Cleopatra kept a small box that could be filled with bees to press against her genitals for stimulation. In recent times, men seem to be the primary formicophiles, such as the fellow who would remove the wings of a fly, submerge himself in bathwater, and then allow the insect to wander around the island formed by his penis (giving a whole new meaning to "check your fly").

While Cleopatra and the bath guy didn't seem to rely, at least explicitly, on fear or disgust (although bees against one's vulva and flies on one's penis might qualify in these regards), others have been more clearly reliant on these emotions. One such case is the man who would lie naked in his backyard and become sexually aroused when ants began to bite. Another is the fellow who would capture bees and hold them against his penis until they stung (there is a different paraphilia in which men generate grotesquely swollen genitals by using bee stings).

Not all paraphilias are autoerotic, however. Individuals also use insects to arouse their partners. In arachnephilia, which may be combined with bondage, a person puts a spider near or on a partner. While a candle on the nightstand might create the right mood for some of us, others find a tarantula in a jar more arousing. Perhaps this is the subtext of the gripping scene in *Dr. No* (1962) in which a spider crawls across James Bond's naked chest in bed—the scene climaxes with his violently crushing the arachnid. And speaking of crushing . . .

* * *

> I must look like a tiny, squirmy silverfish or an oversized white worm or maggot. . . . "Look, there on the floor, you guys, it's my boyfriend. I know it looks like a strange insect but it's him." [a friend makes to fetch a tissue] "Why bother," says Rei. "Let's just step on him!"[56]

Figure 5.3
A tarantula in the genus *Avicularia* was used in the filming of the scene in *Dr. No*. Sean Connery (the best James Bond, as everyone knows) was not in any physical danger. However, some individuals use spiders during foreplay as a means of heightening physiological arousal, which then transfers into improved sexual performance (image by Psychonaught through Wikimedia Commons).

In this passage from the *American Journal of the Crush-Freaks*, the author imagines himself insect sized, and his sexual fantasy culminates in his being ground into the carpet. The paraphilia of crush fetishism involves the desire to be squashed by a woman. Although being trampled to death by a powerful femme fatale would be the epitome of fulfillment, dying is a high price to pay for great sex. So the crush fetishist substitutes another living organism, watches it being crushed (typically with high heels or bare feet), and fantasizes about being the worthless, repulsive, vulnerable, squishy victim. The fetish has been anecdotally traced to a childhood experience of seeing one's mother crushing an insect, which becomes seared into the psyche of the boy (only males exhibit this paraphilia): "In that blink of a wide-open eye, something gets made forever, something gets lost forever."[57]

Until the early 1990s, few people were aware that some men derived sexual satisfaction from watching creatures being crushed by women. A pair of art films produced by Jeff Vilencia put this paraphilia on the avant-garde radar, although *Squish* and *Smush* did not draw much public attention.[58] However, Vilencia went on to make a series of fifty-six crush videos that crossed the line from art to pornography. The plot conventions mirrored those of low-budget porn: an ordinary setting in which a woman happens to encounter the fetishist, followed by a large sheet of white paper on which worms and insects are scattered, and then a bit of contrived dialogue with lots of crushing, including graphic close-ups (e.g., a cricket's head protruding from beneath a shoe).

Valencia's videos were produced for the "soft crush" folks who are satisfied by invertebrates (worms, snails, crickets, grasshoppers). However, not every spineless creature is fair game. According to one "crush mistress," "I do not step on the little spindly legged spiders 'cause they're my friends. But when it comes to bugs, I mean they're just icky little creatures so I can't think why they shouldn't be stepped on!"[59] While some draw the line at spiders, others are not so restrained. The far less common "hard crush" sorts opt for vertebrates (mice, guinea pigs, and kittens). However, even soft-core crush videos were sufficient to raise the ire of the public—and politicians.

When people learned of this paraphilia, the outrage led to congressional hearings at which opponents testified—without offering evidence—that crushing insects is a "gateway fetish" leading to spousal abuse, school shootings, and even the crushing of children. Vilencia became a remarkably articulate advocate for his fellow fetishists. As a vegan and animal rights advocate, he argued that society is profoundly hypocritical. Why is it acceptable to raise sentient mammals in the appalling misery of factory farms to satisfy carnivorous desires but immoral to squash insects to satisfy sexual desires?

Good question, but nobody was willing to go there. House Resolution 1887 sailed through Congress. In a nod to the First Amendment, Bill Clinton signed it with the caveat that the Justice Department apply the law only to "wanton cruelty to animals designed to appeal to a prurient interest in sex."[60] And while

Clinton had a thing for young interns, other presidents seem to have known a thing or two about crush fantasies.

Susan Miller manages to unite insects, disgust, sex, crushing, and politics in a compelling condemnation. In reference to Richard Nixon's ordering his minions to break into the office of Daniel Ellsberg's psychiatrist in search of damning information, she writes: "The President disregards a less powerful man's right to privacy; he chooses to intrude in the man's private sphere and make it public, humiliating him. He makes the man into a squashed bug whose insides spurt out."[61]

That's quite an image. But then, we are capable of spectacular fantasies. If we can picture a person crushed like a bug, then perhaps we shouldn't be surprised that the infested mind can fabricate visions and sensations of insects that aren't really there.

As we'll see, some imaginings involving insects are even more disturbing and debilitating than a cricket underfoot.

NOTES

1. Graham C. L. Davey and Sarah J. Marzillier, "Disgust and animal phobias," in *Disgust and Its Disorders*, ed. Bunmi O. Olatunji and Dean McKay (Washington, DC: American Psychological Association, 2009).
2. Ibid.
3. Ibid.
4. Sheila R. Woody and Bethany A. Teachman, "Intersection of disgust and fear: Normative and pathological views," *Clinical Psychology: Science and Practice* 7 (2000): 291–311.
5. Ibid.
6. Scott R. Vrana, "The psychophysiology of disgust: Motivation, action, and autonomic support," in Olatunji and McKay, *Disgust and Its Disorders*.
7. Dean McKay and Melanie W. Moretz, "The intersection of disgust and contamination fear," in Olatunji and McKay, *Disgust and Its Disorders*.
8. Davey and Marzillier, "Disgust and animal phobias."
9. Peter J. de Jong and Peter Muris, "Spider phobia interaction of disgust and perceived likelihood of involuntary physical contact," *Journal of Anxiety Disorders* 16 (2002): 51–65.
10. Salvador Dalí, *The Secret Life of Salvador Dalí* (New York: Dial, 1942), 129.
11. Aurel Kolnai, *On Disgust*, ed. Barry Smith and Carolyn Korsmeyer (Chicago: Open Court, 2004), 16–22; William Ian Miller, *The Anatomy of Disgust* (Cambridge, MA: Harvard University Press, 1997), 25–26.
12. Sue Hubbell, *Broadsides from the Other Orders: A Book of Bugs* (New York: Mariner, 1998), 158.
13. Sandra A. N. Mulkens, Peter J. de Jong, and Harald Merckelbach, "Disgust and spider phobia," *Journal of Abnormal Psychology* 105 (1996): 464–68.
14. Jason M. Armfield and Julie K. Mattiske, "Vulnerability representation: The role of perceived dangerousness, uncontrollability, unpredictability and disgustingness in spider fear," *Behaviour Research and Therapy* 34 (1996): 899–909.

15. Jacqueline Ware, Kumud Jain, Ian Burgess, and Graham C. L. Davey, "Disease-avoidance model: Factor analysis of common animal fears," *Behaviour Research and Therapy* 32 (1994): 57–63.
16. Mark van Overveld, Peter J. de Jong, and Madelon L. Peters, "Differential UCS expectancy bias in spider fearful individuals: Evidence toward an association between spiders and disgust-relevant outcomes," *Journal of Behavior Therapy and Experimental Psychiatry* 37 (2006): 60–72.
17. Kolnai, *On Disgust*, 29–33.
18. Miller, *The Anatomy of Disgust*, xi.
19. My dissonant relationship with these creatures is detailed in my collection of essays *Grasshopper Dreaming: Reflections on Killing and Loving* (Boston: Skinner House, 2002).
20. Susan B. Miller, *Disgust: The Gatekeeper Emotion* (Hillsdale, NJ: Analytic Press, 2004), 10.
21. Paul Rozin, Larry Hammer, Harriet Oster, Talia Horowitz, and Veronica Marmora, "The child's conception of food: Differentiation of categories of rejected substances in the 1.4 to 5 year age range," *Appetite* 7 (1986): 141–51.
22. Miller, *Disgust: The Gatekeeper Emotion*, 4–5.
23. April E. Fallon, Paul Rozin, and Patricia Pliner, "The child's conception of food: The development of food rejections with special reference to disgust and contamination sensitivity," *Child Development* 55 (1984): 566–75.
24. Miller, *The Anatomy of Disgust*, 12–14; Craig N. Sawchuck, "The acquisition and maintenance of disgust: Developmental and learning perspectives," in Olatunji and McKay, *Disgust and Its Disorders*.
25. Sawchuck, "The acquisition and maintenance of disgust."
26. Paul Rozin, Maureen Markwith, and Carol Nemeroff, "Magical contagion beliefs and fear of AIDS," *Journal of Applied Social Psychology* 22 (1992): 1081–92.
27. Sawchuck, "The acquisition and maintenance of disgust."
28. Paul Rozin, Jonathan Haidt, and Clark R. McCauley, "Disgust: The body and soul emotion," in *Handbook of Cognition and Emotion*, ed. Tim Dalgleish and Mick Power (Chichester, England: Wiley, 1999).
29. Valerie Curtis and Adam Biran, "Dirt, disgust, and disease: Is hygiene in our genes?," *Perspectives in Biology and Medicine* 44 (2001): 17–31; Graham C. L. Davey, Angus S. McDonald, Uma Hirisave, G. G. Prabhu, Saburo Iwawaki, Ching Im Jim, Harald Merckelbach, Peter J. de Jong, Patrick W. L. Leung, and Bradley C. Reimann, "A cross-cultural study of animal fears," *Behaviour Research and Therapy* 36 (1998): 735–50.
30. Miller, *Disgust: The Gatekeeper Emotion*, 191.
31. Ibid., 13.
32. Curtis and Biran, "Dirt, disgust, and disease."
33. Miller, *The Anatomy of Disgust*, 64.
34. Sawchuck, "The acquisition and maintenance of disgust."
35. Miller, *Disgust: The Gatekeeper Emotion*, 4; Rozin et al., "Disgust."
36. Miller, *The Anatomy of Disgust*, 11.
37. Julieta Ramos-Elorduy, "Anthropo-entomophagy: Cultures, evolution and sustainability," *Entomological Research* 39 (2009): 271–88; Peter Menzel and Faith D'Aluisio, *Man Eating Bugs: The Art and Science of Eating Insects* (Berkeley, CA: Ten Speed Press, 1991).
38. Michael Lewis, *Shame: The Exposed Self* (New York: Free Press, 1992), 206.
39. Brian Morris, *Insects and Human Life* (Oxford: Berg, 2006).

40. Lucien Malson, *Wolf Children*, trans. Edmund Fawcett, Peter Ayrton, and Joan White (New York: Monthly Review Press, 1972).
41. Miller, *Disgust: The Gatekeeper Emotion*, 49.
42. Miller, *The Anatomy of Disgust*, 16.
43. Ibid., 17.
44. Daniel Kelly, *Yuck! The Nature and Moral Significance of Disgust* (Cambridge, MA: Bradford/MIT, 2011).
45. Miller, *The Anatomy of Disgust*, 16.
46. David H. Barlow, *Anxiety and Its Disorders: The Nature and Treatment of Anxiety and Panic*, 2nd ed. (New York: Guilford Press, 2002).
47. Brenda Love, *Encyclopedia of Unusual Sex Practices* (New York: Barricade Books, 1992).
48. Ratnin Dewaraja, "Formicophilia, an unusual paraphilia, treated with counseling and behavior therapy," *American Journal of Psychotherapy* 41 (1987): 593–97; Ratnin Dewaraja and John Money, "Transcultural sexology: Formicophilia, a newly named paraphilia in a young Buddhist male," *Journal of Sex & Marital Therapy* 12 (1986): 139–45.
49. Dewaraja and Money, "Transcultural sexology," 141.
50. Ibid.
51. Ibid., 142.
52. Ibid.
53. Dewaraja, "Formicophilia," 594.
54. Love, *Encyclopedia of Unusual Sex Practices*, 301.
55. Ibid., 301–2.
56. Hugh Raffles, *Insectopedia* (New York: Pantheon, 2010), 270.
57. Ibid., 280.
58. The following story of Vilencia and the political response to his videos is adapted from Raffles, *Insectopedia*, 267–90.
59. Ibid., 269.
60. Ibid., 288.
61. Miller, *Disgust: The Gatekeeper Emotion*, 20.

CHAPTER 6

The Terrible Trio: Imagining Insects into Our Lives

Minds are snarls of perceptions and beliefs—and describing what happens inside our heads is fraught with imprecision. One of the most tangled sets of experiences consists of hallucinations, delusions, and illusions. We'll take these one at a time, sort of. It would be handy if the infested mind reflected the orderly categories that science has cut out of our inner world, but the three threads that produce a knot of mistaken thoughts don't always pull apart neatly. Even so, we'll begin by pulling on the thread of hallucination.

HALLUCINATIONS: THEY'RE ALL IN YOUR MIND

A hallucination is an internally originating perception, such that the experience is uniquely the hallucinator's own. Individuals see nonexistent insects and spiders in rather diverse contexts, ranging from geriatric centers to hostage situations.[1] However, the classic entomological hallucination is experienced by alcoholics. Only about 5 percent of people suffer delirium tremens during alcohol withdrawal, but a quarter of them hallucinate. These people often "see" insects on the walls or on their bodies—a phenomenon termed *zoopsia* (*zo-* meaning animal and *-opsis* meaning seeing).[2] Zoopsia is often paired with tactile hallucinations—the itchy-tingling feeling that the insects are crawling across the skin, which is called *formication* (as with formicophilia, this term is derived from the scientific name for ants).

Along with alcohol withdrawal, chronic use of some drugs can generate hallucinations. So common are these perceptions that they have been named "cocaine bugs" and "crank bugs" (the latter in association with long-term amphetamine use).[3] Formication is the most common hallucination, although sometimes the nonexistent insects are also seen. One of the earliest descriptions of "cocaine bugs" was provided by Ernst von Fleischl-Marxow, an Austrian physician and a friend of Sigmund Freud—who advised Freud to use cocaine.[4]

Other drugs also have the capacity to induce formication, although when it comes to the personal nature of itchiness, it is difficult to know whether this is an internally originating misperception or a real sensation. In any case, medications such as phenelzine and corticosteroids can induce tactile hallucinations. Various ailments can also do so, including cancers (e.g., Hodgkin's lymphoma and chronic lymphocytic leukemia), infections (e.g., tuberculosis and syphilis), metabolic anomalies (e.g., renal dysfunction and hyperthyroidism), neurological disorders (e.g., Huntington's chorea and Parkinson's disease), and malnutrition (e.g., vitamin B_{12} and B_3 deficiencies and iron deficiency).[5]

ILLUSIONS: THEY FEEL REAL

Like a hallucination, an illusion is a misperception. However, illusions involve actual sensory stimuli (usually taken to originate outside the body, so drug-induced misperceptions are classified as hallucinations). Under certain conditions, people perceive objects and events in mistaken ways, and illusions are often shared among those in the same setting. For example, most of us are duped by optical illusions, mirages are typically witnessed by people standing in the same place, and the entire audience is fooled by an illusionist turning a silk scarf into a butterfly.

In illusory parasitosis, the individual experiences a genuine stimulus but incorrectly attributes the sensation to an insect or other parasite. While working in the field, I commonly mistake the sensation of grass against my legs as being an insect clinging to me. And on that long drive home after the encounter with the swarm, I could not shake the illusion that every itch or tingle was a grasshopper under my shirt or in my hair.

Although it seems unexpected that a scientific setting would be ripe for illusory parasitosis, a classic case unfolded in 1967 in the physics laboratory of a national research agency.[6] Ten women working at the facility reported prickling and tingling sensations. An exterminator was brought in and applied an arsenal of pesticides, including DDT, chlordane, fenthion (an organophosphate), warfarin (an anticoagulant to kill any rogue rodents), and mothballs for good measure.

Despite the best efforts of the pest control company, the employees continued to feel small creatures crawling on their legs, scalps, and necks. One victim reported that her friends who worked for the telephone company often suffered from cable mites—and suggested that maybe these creatures were the problem.

A public health entomologist and a veterinarian were brought in and soon eliminated arthropods as suspects. Although there are no such things as cable mites (more about these illusory arthropods in a moment), the women were not imagining their skin irritation. With some CSI-like sleuthing, the consultants pieced together the story. Some weeks earlier, the laboratory had purchased equipment that required an overhead power cable. During installation of the cable, particles of rock wool (similar to fiberglass insulation) in the ceiling broke loose and were drawn into the ventilation system, from which they were showered slowly over the cubicles where the women were working. These tiny fibers are nearly invisible, but they itch like the dickens. When the laboratory and ventilation system were thoroughly cleaned, the "infestation" disappeared.

The imaginary creatures associated with illusory parasitosis have been variously called cable mites, cable bugs, cable fleas, cable lice, and paper lice. The term *cable bug* was coined by British telephone switchboard operators who were prone to skin irritations that they interpreted as bites inflicted by tiny creatures living in the masses of cables in their workplaces.[7] Industrial hygienists found that synthetic fabrics, which were all the rage at that time in women's clothing, had a tendency to become electrostatically charged—and repeated discharging across the skin could induce small

Figure 6.1
A typical work setting for keypunch operators in the 1960s shows the combination of factors that could foster illusory parasitosis. Chads falling from the punch cards would cling to synthetic fibers of the women's clothing and cause irritation. Sedentary work on a repetitive task in a drab, stuffy, crowded room is a psychological recipe for imagined infestations (image by Arnold Reinhold through Creative Commons).

lesions resembling insect bites.[8] So "polyester bugs" might have come closer to naming the culprit.

Phillip Weinstein, a professor of public and environmental health, warns that we must be very careful to exclude the possibility of a genuine insect infestation before asserting that people are afflicted with illusory parasitosis.[9] Cable mites are mythical, but various arthropods can infest homes and workplaces. Weinstein gives an example from his own laboratory in which he and others were plagued by an undetectable assailant. Frustrated by the perception of being bitten for many days, he eventually nailed the mysterious attacker. Outside his laboratory, pigeons and starlings had set up house and brought with them a colony of mites. Each smaller than the period at the end of this sentence, these eight-legged parasites prefer feathered hosts but aren't too picky when hunger drives them indoors.

* * *

In Dante's *Inferno*, we learn that itching can be hell. Consider the punishment for "falsifiers" (alchemists, counterfeiters, imposters, and liars):

> Then I saw these two scratch themselves with nails
> Over and over because of the burning rage
> Of the fierce itching which nothing could relieve.

If itching is infernal and parasites are diabolical, then what sort of fiendish psychology gives rise to the torment of illusory parasitosis? The answer is a concatenation of our how our minds have been shaped by evolutionary history, the intellectual capacity for inference, and our experiences (both personal and social).

Our primate relatives spend a great deal of time grooming—a polite term for removing ectoparasites from themselves and one another, which also serves a social function, like playing bridge or taking coffee breaks. Some species even have evolved specialized claws and teeth to facilitate parasite removal, and social primates spend up to 20 percent of their time grooming.[10] All of this suggests that for a very long time our ancestors were plagued by ticks, lice, and other vermin.

Itching evolved as a means of alerting us to something having settled on our bodies, and the compulsion to scratch protected us from imminent danger. When a mosquito lands and feeds, the result might be malaria; a fly could bring sleeping sickness; a louse, typhus; a flea, bubonic plague; a tick, Crimean-Congo fever; a mite, scrub typhus. There were clear advantages to our ancestors who responded to the itch of an arthropod invader with vigorous scratching. In fact, recent studies have demonstrated that the fine body hair on humans continues to play a role in our ability to detect ectoparasites—researchers who

Figure 6.2
An olive baboon adult grooming a juvenile provides social bonding through physical contact and hygiene by removing vermin and other irritants. Being highly attuned to itchiness meant that our ancestors detected arthropods that transmitted deadly pathogens. We have inherited their proclivity for itchiness and intensive skin and hair grooming, as seen in modern teenagers (image by Muhammad Mahdi Karim through Wikimedia).

placed bed bugs on the shaved and unshaved arms of subjects found that hairy individuals were two to three times more sensitive to the presence of the insects than were people whose arms had been shaved.[11]

So primed are we for itching and scratching that we can make ourselves itchy just by thinking about tiny creatures wandering across our skin or through our hair. This proclivity was demonstrated by a German professor of psychology who gave a lecture in which he showed itch-inducing images (e.g., fleas and lice) for the first part of his talk and neutral images for the latter half (e.g., baby skin and bathers). The audience performed as expected, with a markedly higher frequency of scratching during the initial slides.[12] In fact, take a moment to think about whether you've felt a bit itchy while reading about all of this. In the course of composing this chapter, I found myself intermittently scratching during every writing session.

Humans are also highly predisposed to find cause-and-effect associations. We are skilled at recognizing regularities, and it is highly adaptive for us to be able to infer what was responsible when something good or bad happens so that we can either seek or avoid the preceding conditions in the future. But how we apprehend the world in these terms is not at all as we might expect.

It would make sense that we directly perceive the outside world, including the presence of a scratch-inducing louse. However, neuroscientists have discovered that perceptions are not so straightforward. Rather than an unmediated grasp of reality, perception is the brain's "best guess" about what is happening. The mind integrates sensory input, individual experience, and innate tendencies to construct a theory of the world (e.g., "I'm infested"). In short, perception is inference.[13]

When it comes to the sensation of itching, our brains assemble a coherent account of what is causing this state of affairs. And if the condition is chronic, our cognitive abilities provide us with the means to construct a reasonable explanation of what is happening. Historically, it was quite plausible for humans to figure that persistent itching was caused by an infestation—and lots of scratching was the way to deal with a family of lice having taken up residence. In today's world, however, this belief is likely to be false.

* * *

The hallmarks of illusory parasitosis are the ways in which the disorder arises and spreads. Transposition and transmission are the twin pillars of establishing the mistaken perception in one or a few individuals—and then spreading it to dozens of others.[14] Few psychological conditions have the capacity to generate such remarkable outbreaks.

In the process of transposition an individual becomes itchy and then, in a search to explain the sensation, blames nonexistent arthropods. This is like seeing a silk handkerchief apparently turn into a butterfly and crediting the performer with magical powers. With illusory parasitosis, affected individuals often have some background experience, such as a flea infestation at home (analogous to the magician's banter that sets you up for a misperception), along with an itchy sensation. The problem (I'm itchy) is transposed from a past setting (home) to the present one (work, where there may be both physical discomfort and psychological stress).

As for transmission, our suggestibility with regard to itchiness means that both sensation and (mis)perception are highly contagious. Indeed, the first cases of illusory parasitosis were described as "mass entomophobia." Although the condition was not actually a phobia, the impression of collective hysteria was vividly evoked by this term. However (to fairly reflect the complexity of the infested mind), in some cases of illusory parasitosis unaffected individuals develop a fear, even a transitory phobia, of catching whatever is afflicting the others. And, as with phobias, women are far more prone than men to collective misperceptions of insects crawling on them.[15]

So when an employer has dozens of workers scratching and complaining about nonexistent insects, what can be done?

* * *

In the 1960s—the heyday of illusory parasitosis in the workplace—a common tactic was to call in an exterminator to conspicuously treat the problem area. Employers were often complicit in the administration of this psychological placebo, which was sometimes just a chemical deodorant.[16] More often, a nonpersistent insecticide was used. The rationale of the exterminators seemed to be something like, "There *could* be an actual infestation that I overlooked, so let's treat and see if it works." And the act of having done something tended to quell complaints.

Today, it is a violation of federal law to apply an insecticide unless pests are present—and there are no chemicals approved for imaginary insects. An exterminator might spray water or deodorant, but this could constitute fraud if the client is not informed. And in any event, entomologists generally maintain that it is unethical to treat a nonexistent infestation.[17] Deceit, even with the best of intentions, is morally problematic. At least if such a tactic is pursued to "treat" illusory parasitosis, it seems best left to those trained in psychology. Entomologists diagnosing mental illness and employing therapies makes about as much sense as psychologists examining fields and calling in crop dusters. Fortunately, industrial psychologists figured out some keen approaches to exterminating illusory insects.

A two-pronged approach proved most effective. First and most obviously, finding and removing the source of the illusion was important. Eliminating the irritation caused by fibers or particles from carpets, fabrics, paper, cardboard, or insulation was sometimes sufficient. A case involving keypunch operators who complained of being bitten was solved with the removal of a fan that had been blowing chads from the punch cards onto the legs of the women.[18] However, in an ironic twist, some solutions of this sort made imaginary infestations real.

Often, the illusion of parasitosis was associated with electrostatic charges, and a common approach to this problem was to increase the humidity. High humidity, however, can foster the growth of minute fungi on paper and cardboard, which feed microscopic mites. The tiny interlopers are visually undetectable, but they can be felt walking across the skin of people who handle the infested material—producing "office worker's itch." To make matters worse, the mites are prolific poopers, and their feces can induce an allergic rash.[19]

Even when modifying the physical environment successfully eliminated the stimulus, modifications in the social environment were sometimes needed. Psychologists ascertained that working conditions contributed to the emergence and spread of illusory parasitosis. In particular, the malady tended to appear when crowded, sedentary workers performed repetitive clerical work under intense pressure in drab, cluttered surroundings. In short, aesthetically and intellectually dull conditions fostered outbreaks. In one well-documented case, eight women working in a dreary, stuffy, windowless room complained of being bitten by some sort of insect. When the office was painted, ventilated, rearranged, and decorated, the complaints disappeared.[20] In other instances, addressing workplace stress was important.

The most massive case of illusory parasitosis required a form of integrated imaginary-pest management.[21] Female employees in an airline reservation office began to complain of tingling sensations around their ankles. As others became afflicted with itchiness, the management called in a pest control operator, who refused to treat in the absence of any direct evidence of insects. An entomologist from the Los Angeles County Health Department discovered that a telephone cable running under the desks in the office was generating a static charge. This caused the sensation that insects were crawling on the women's legs, so shielding was added to the cable. Having paid attention to the work of industrial psychologists, the fellow also observed that morale was poor among the staff. The women were working under considerable pressure and were being intensively monitored by supervisors who sat in a darkened booth at the front of the room. As a part of the holistic treatment program, the light in the supervisors' booth was kept on to reduce the workers' feelings of powerlessness that came with being constantly scrutinized by unseen overseers (perhaps the high frequency of illusory parasitosis in women reflected a gender-based difference in power along with a biological predisposition). Between covering the cable and uncovering the supervisors, the outbreak was quashed.

And nipping hallucinations and illusions in the bud can be terribly important, as imaginary insects can become embedded in the infested mind—with devastating consequences.

DELUSIONS: I KNOW THEY'RE THERE

A delusion entails having an erroneous and persistent *belief*, not just a mistaken perception. A false belief can arise from an illusion—a person attending a magic show might come to grandiosely believe that she shares the illusionist's power to make butterflies appear. Or perhaps the magician took a round of corticosteroids and developed a tactile hallucination of itchiness that gave rise to a persistent delusion that he was infested with caterpillars under his skin. With this move from misperception to false belief, we have one of the more tragic forms of an infested mind: delusory parasitosis.

Perhaps the most famous case of delusory parasitosis is that of Salvador Dalí.[22] The artist could induce hallucinatory states by a process he described as "paranoiac critical" and then use the subconscious images in his paintings. Perhaps if I had maintained and mastered the practice of lucid dreaming, I wouldn't have been psychologically overwhelmed in the midst of a maelstrom of grasshoppers. But perhaps being a reasonably well-adjusted entomologist has some advantages over being a profoundly disturbed artist. For Dalí, the images of grotesque human forms swirled with those of insects. And his paintings reveal his fears and fantasies, which were often rooted in his youth.

The artist's childhood experiences with insects entrenched his entomophobia and evidently gave rise to delusory parasitosis in adulthood. The Spanish filmmaker Luis Buñuel recounted a visit with Dalí:

> I found him stripped to the waist, an enormous bandage on his back. Apparently he thought he'd felt a "flea" or some other strange beast and had attacked his back with a razor blade. Bleeding profusely, he got the hotel manager to call a doctor, only to discover that the "flea" was in reality a pimple.[23]

According to Dalí, the blemish was a birthmark, and he mistook it for a bed bug or cockroach or tick,[24] but given the artist's proclivity for muddling memories, it is hard to know which account is more accurate. In any case, his delusion led to a spectacular act of self-mutilation.

A somewhat more typical but no less traumatic story is vividly captured in a 1954 account of delusory parasitosis in an elderly couple.[25] They reported that their house was infested and that there were insects crawling on them. Ashamed of their condition, they withdrew socially and the delusions worsened as "thousands of tiny, pin-point specks that glitter like diamonds" hatched into curled-up larvae that developed into little brown bugs that inflicted insufferable itching.

In an effort to rid themselves of these imaginary insects, the couple boiled their bedding and clothing daily, baked their pillows and cushions, burned their couch and bed, and sprayed their furnishings with a mixture of carbolic acid, hot water, blue vitriol, rubbing alcohol, and DDT. They subjected themselves to three scalding baths a day, and before going to bed the wife sprayed her husband with a weak mixture of carbolic acid and DDT. In a downward spiral, the delusions combined with sleeplessness (from formication) to trigger hallucinations—"the bugs came down the wall, marched in single file across the room, ran up my leg under my underwear, and came out my neck and disappeared"[26]—which solidified the wildly mistaken beliefs.

Such wretched tales appear in the scientific literature as far back as the nineteenth century. In 1894, Georges Thibierge reported "acarophobes"—people who were pathologically afraid of contracting scabies. Two years later, Léon Perrin coined the unwieldy term "parasitophobic neurodermatitis" to describe an intense fear or mistaken impression of parasites living under the skin.[27]

In 1930, Karl Axel Ekbom published his classic paper in which he clearly differentiated entomological delusions from entomophobia based on seven case studies.[28] The key was that his patients believed they were infested, which was fundamentally different from the fear of becoming infested. Thus was born Ekbom's syndrome, named in honor of the Swedish neurologist, who went on to identify another debilitating condition—restless leg syndrome. The term *delusion of parasitosis* was proposed in 1946 and later shorted to *delusory parasitosis*.[29]

Delusory parasitosis may not be the whole story for some patients, as it is often confounded with other mental disorders. This was the case with a forty-one-year-old man who complained of feeling persecuted, hearing voices, and having flies crawling on his face. The poor fellow would continuously rub his face to keep the insects away, and he was eventually hospitalized. Long-term treatment with antipsychotic medications helped to control his persecutory delusions and auditory hallucinations, but he could not stop rubbing his face.[30]

* * *

Although delusory parasitosis manifests in various ways, there is a core of four symptoms that signal this condition. First is the "matchbox sign," which was first described in a leading medical journal in 1983.[31] This ominous symptom is named for the proclivity of the sufferer to present a matchbox containing the offending insects. Microscopic examination of the contents of the matchbox reveals lint, scabs, fibers, and household detritus, but no arthropods, not even a forlorn dust mite. Typically, the failure of the professional—usually a dermatologist or entomologist—to find any evidence of an infestation leads to increasingly intensive collections and impassioned presentations. My conversations with colleagues in pest management suggest that we might update this symptom as the "ziplock sign." Extension entomologists cringe when a person arrives with a baggie full of dusty tissues and grit-speckled cellophane tape.

Next, the patient frequently offers vivid accounts of the insects and their origins.[32] Typically the insects are described as black or white. They tend to jump when prodded, as would be expected of an electrostatically charged crumb. The precise description depends on the individual's background, with commonly imagined culprits being fleas, lice, mites, and bed bugs among the more biologically literate. Those with less knowledge more often refer simply to "bugs" or even "black things." Sometimes the source of the infestation is said to be another person or an animal, particularly a scruffy dog or mangy cat. Surprisingly often, the sufferer claims the insects came from a common household product such as toothpaste, petroleum jelly, or cosmetics.

Third, the individual's belief is unshakable. This hallmark tenacity is exemplified by the case of a zoologist who could not be convinced that the gunk she brought in did not include arthropods, insisting that the microscope was wrong.[33] Often the patient provides such a lucid account of the infestation that family members come to adamantly support, and thereby reinforce, the delusion.

Finally, sufferers have often done significant damage to their skin, either through incessant scratching or with chemical treatments.[34] When fingernails aren't enough, people resort to using tweezers, knives, and razor blades (as in the case of Dalí) to dig the insects from beneath their skin—and when shallow abrasions don't do the trick, they deepen the wounds. Some sufferers turn to a horrific range of chemical substances: acids, alcohols, alkalis, astringents, balms, cleaning compounds, detergents, gasoline, insecticides (from DDT to dog dip), kerosene, oils, ointments, shampoos, and soaps.

Not only are those suffering from delusory parasitosis a hazard to themselves, but they may also endanger others. In one case, a British man complained repeatedly to the local housing council about an infestation.[35] The remarkably indulgent officials moved him three times after he burned and flooded apartments to rid himself of insects, and they fumigated his residences on eight occasions—such is the persuasive power of the delusional individual to convince others of a nonreality. Upon setting two more fires in 1986, he was admitted to the hospital, where he "shouted loudly and aggressively about his detention and complained that 'insects swarm all over me.'"[36] After antipsychotics calmed him down, he returned home and switched from using fire to applying large quantities of insecticides, which evidently did not trigger further interventions by officials.

A more frightening case unfolded in 1967, when a fifty-eight-year-old British woman tried to shoot her family doctor with a hunting rifle.[37] The woman suffered from delusory parasitosis and provided her physician with "specimens" in a matchbox. When he would not affirm her beliefs, she became convinced that he had engineered a plot to infest her. It didn't help that her alcoholic husband reinforced her paranoia and parasitic delusions. The poor woman was institutionalized for twenty-four years, being occasionally released only to attempt suicide each time. Some cases end even more tragically—a California woman suffering from delusional parasitosis killed her doctor after he advised her to visit a psychiatrist.[38]

While such behavior seems extreme, the maddening effect of being absolutely inundated by biting insects is a common theme in writings about presumably sane people traveling into regions filled with black flies and mosquitoes.[39] Stories of frontiersmen driven mad by biting insects in Alaskan tundra, the Canadian marshlands, and Minnesota's Boundary Waters are echoed in the journals of modern adventurers:

> The black flies are the worse that I have ever seen. There must be a hatch coming off the water at that very moment. I've heard that sometimes certain animals, especially caribou on the tundra, can be driven insane by the torment of the black flies. I believe it. I suppress an almost overwhelming urge to run and escape.[40]

Accounts of crazed caribou—as well as cows, horses, and moose—suggest that the torment can trigger extraordinarily bizarre behavior even in other species.

* * *

Strictly speaking, delusory parasitosis is the mistaken belief that one's body is infested, so what if a person is convinced that his or her home is infested? A too-clever entomologist called this condition "delusory cleptoparasitosis."[41] A cleptoparasite steals resources that its host has collected or stored. Some of the best-known cases of cleptoparasitism involve a bee or wasp infesting the home of another species by laying eggs in its nest. The entomologist's term hasn't caught on, although those who suffer from delusory parasitosis often believe that their surroundings are infested.

Perhaps the most important feature of delusory cleptoparasitosis is that this condition can precede the onset of delusory parasitosis. In one case, a man complained that his home was overrun with wood-feeding insects.[42] He told an exterminator that picture frames, cabinets, and boxes were infested, but the exterminator could see no evidence of wood-destroying insects. Several weeks later, the man and his housemate began to complain that insects were now burrowing into their skin, but the exterminator was still unable to find any insects and refused to spray. When the men badgered the company with increasingly belligerent demands, the police put an end to the harassment.

* * *

The mistaken belief that defines delusory parasitosis has been found to originate in complex and varied ways. Take the case of a thirty-four-year-old woman reported in the journal *Plastic and Reconstructive Surgery*.[43] She was treated for a deep ulceration in her chin that extended to the bone. When she returned a few weeks later for treatment of another lesion—along with scratches and abrasions on her legs and trunk—the doctor determined that she was experiencing delusions arising from methamphetamine use. The self-inflicted damage was the result of her conviction that insects had laid eggs that hatched under her skin. In this instance, there was a happy ending thanks to drug rehabilitation.

This case reveals a common path from hallucination or illusion to delusion. What began as a drug-induced formication served as the basis for the woman's conviction that she was infested—a belief that was firm enough to engender self-mutilation. So a key to understanding the nature of the complex psychological processes that culminate in delusory parasitosis is recognizing that our real experiences are woven into our unreal beliefs. Often a patient has had some previous disturbing encounter with insects, such as an infestation of cockroaches, fleas, or lice.[44] Then the person becomes hypervigilant, and every little tickle or itch becomes a tactile illusion—the first step onto a slippery slope toward delusion.

In a kind of perversely self-fulfilling world of sensation, perception, and imagination, the individual begins to scratch in an effort to remove the invader. What dermatologists call the itch-scratch cycle is set in motion.[45] Scratching provides momentary relief, followed by intensified itchiness. The delusion begins to take hold and deepen as persistent scratching (or chemical treatment) leads to damage and infections that exacerbate the feeling of being infested. The individual has managed to create an actual sensation that is interpreted as an infestation.

Thanks to Sigmund Freud there is a rather more bizarre and intriguing explanation for how delusory parasitosis develops. Some psychologists maintain that this condition originates from a woman's body image, which combines with sexual guilt to instill a fear of superego punishment. The notion here is that women have "unconscious sexual guilt with an attempt to ward off feared, and at the same time, hoped for, invasion of a sexual nature. The problem of intromission by the male organ may be symbolically displaced upon insects which cannot be eradicated, just as the guilt-laden sexual impulses cannot be eradicated."[46] So the delusion arises from shame and a fear of punishment by the stodgy superego for the woman's having had forbidden sex. In other words, imaginary insects are the minions of internalized prudish parents. As nutty as this sounds, Freudians offer some intriguing evidence.

Women suffering from delusory parasitosis not infrequently claim that the bugs originated in cosmetics or other skin products, hence the connection to a poor body image. And in some cases sexual guilt seems to foster the delusion, as in the case of a forty-year-old divorcee who contracted crab lice during an extramarital affair and later developed delusory parasitosis.[47] Or consider the thirty-eight-year-old widow who'd had extramarital affairs and told her doctor that when bathing after sex she could see black specks emerging from her back.[48]

If the origins of delusory parasitosis are complicated, at least some of the demographic patterns of the disorder are fairly clear. The typical delusional patient is a sixty-year-old (almost all are older than fifty) woman (suggesting a connection with entomophobia and illusory parasitosis) of average intelligence (able to infer effectively and argue persuasively).[49] The contagious nature of the delusional state, along with any concomitant illusional perceptions, means that others may be afflicted. In 10 to 25 percent of the cases, the delusions are shared by family members in a phenomenon called *folie partagée*, meaning "shared madness."[50] One study of 282 cases revealed 24 instances of *folie à deux* (two household members were afflicted) and three cases of *folie à trois* (three members).[51]

Delusory parasitosis is the most common of the somatic delusional disorders, which involve beliefs about one's body as opposed to delusions of love, grandeur, jealousy, and persecution.[52] As a whole, delusional disorders are reportedly rare in the United States, with a prevalence of around 0.03 percent,[53] meaning that perhaps one hundred thousand people suffer from some form of these disorders. However, many psychologists believe that delusory parasitosis is far more common than the literature would suggest.[54] Virtually every dermatologist has encountered cases, and most extension entomologists and pest control operators know of people who appear to have the disorder. A former graduate student of mine who became a manager in one of the largest pest management companies on the East Coast said that it was not unusual to encounter several potential cases in a week—either directly through his contact with the public or through referrals from those he supervised.[55]

* * *

We can argue about how many people are afflicted with delusory parasitosis, but there seems to be no debate that the condition is debilitating, is unlikely to resolve itself, and requires some form of intervention. However, just how assistance should be provided is not always clear. The difficulty begins with the individual's initial contact—often with an entomologist or pest control operator. What to tell an individual presenting with the symptoms of delusory parasitosis is a tough call. On one hand, challenging the false belief often prompts the individual to either work harder to acquire evidence or seek a second opinion (or third or fourth or . . .). Conversely, responding vaguely and sympathetically runs the risk of deepening the delusion and making psychological treatment—if the individual ever receives help—all the more difficult.

The general recommendation is for the insect expert to state firmly that no arthropods are present when such is the case.[56] This tactic might nip a still-forming delusion in the bud, but the rationale is one of principle rather than pragmatics. After all, not long ago lying was considered a viable tactic—and there was evidence that it worked.

Remember that elderly couple who boiled their sheets, baked their pillows, and sprayed carbolic acid laced with DDT? Well, back in 1954 a dose of deceit was apparently worth a try. The entomologist knew there were no insects but told the couple that a recently discovered chemical would eliminate their infestation. He provided them with a small amount of DDT and detailed instructions for spraying their home over the course of a month. When the old folks completed the protocol, they were delighted. They gratefully reported that the bugs were completely eliminated. Finally able to sleep through the night, they regained their health and went back to attending church and hosting visitors.

Other reports from the 1950s affirm that placebos could cure delusory parasitosis. But as with such treatments for illusory parasitosis in the 1960s, there were evidently ethical and legal concerns. In most placebo treatments for delusory parasitosis, exterminators applied at least a weak dose of insecticide.[57] It is not clear why they didn't just spread some talcum powder or spray plain water. Perhaps they drew the line at charging a customer for doing nothing—an unnecessary chemical application was one thing, fraud was another. However, in one particularly interesting case the exterminator managed to relieve a customer's delusion without any application at all.

When a woman called in the pest control company to treat her apartment, she began by explaining that inanimate specks on the kitchen walls were the source of the biting insects. Her husband told the technician that his "batty" wife was being treated for a nervous disorder. The exterminator considered the situation and developed a rather unconventional solution. He suggested that the insects might be hiding in the furniture, so the whole apartment would need to be treated. Presumably the woman was relieved that something could be done, but then he quoted her an exorbitant price for the treatment. When her delusions came into conflict with dollars, the latter won out. She deferred treatment and several months later reported that she had not encountered any more insects.[58]

Sometimes placebos provided only temporary relief. When a woman reported that insects were biting her and laying eggs all over the house, the exterminator paid a visit and was presented with jars of bath tissue that supposedly contained the pests. Even after the woman viewed the dust and lint through a microscope, she was unconvinced that no insects existed, so the fellow relented and sprayed her home. For two weeks all was well, and then she called to report frantically that the vermin were back. She was urged to see a doctor, but the outcome of this case is unknown.[59]

Surely it is a good thing that today's exterminators are prohibited from diagnosing mental disorders and using placebos (and it's also good that psychologists don't try to manage termites). So how do modern therapists handle this disorder?

Delusions are difficult to treat, particularly when the patient denies that there is a problem. The bible of psychology, the *Diagnostic and Statistical Manual of Mental Disorders*, offers helpful advice—along with not-so-helpful suggestions. We are told that delusory parasitosis might be a way for the person to handle problems that could manifest in less manageable ways, as if shooting your physician were relatively manageable. Experts in the field also admonish therapists to avoid expressing "amusement, ridicule, and rejection of the patient's important belief,"[60] which makes one wonder whether psychologists really need to be told not to mock their patients.

The basic idea is that the therapist should neither confirm nor challenge the patient's claim of being infested. While gingerly keeping the patient engaged in a kind of nonjudgmental limbo, the therapist can employ one of several options—and avoid those that don't work. As one may suspect, electroconvulsive therapy (ECT) and psychosurgery are bad options.[61] After that, things get a bit less clear. Some reports suggest that psychotherapy and psychoanalysis work if the delusion of parasitosis is rooted in repressed sexual conflicts.[62] Others, however, contend that psychotherapy is no more successful than ECT,[63] although the former is presumably far less traumatic. More recent studies indicate that cognitive behavioral therapy, which entails having the patient systematically reunite emotional feeling with rational thought, is often successful.[64]

Applying chemical insecticides to treat a nonexistent infestation is forbidden, but applying psychoactive chemicals is encouraged. Antipsychotic drugs such as phenothiazine and pimozide have worked with some patients.[65] Phenothiazine has some unusual properties that might provide pathways of relief other than its psychotropic effects. This drug was introduced as an insecticide in the 1930s, and it controls infestations of parasitic worms in livestock.[66] So, if there is a cryptic infestation, phenothiazine could work for rather unexpected reasons. Pimozide appears to be the antipsychotic most commonly prescribed for delusory parasitosis—and it has the bonus effect of relieving itchiness.[67]

* * *

If finding the best therapy is a matter of contention, who applies that therapy is no less controversial. Entomologists should not be playing doctor, but which kind of doctor ought to be treating delusory parasitosis is not a simple matter.

The core of the debate is whether dermatologists or psychologists/psychiatrists should treat delusory parasitosis. At issue is not who is better qualified to treat mental disorders, but whether doing nothing is better than a dermatologist providing psychological therapy. The argument is that if the dermatologist (like the entomologist) directs someone to a mental health professional, then the delusional individual will likely be lost from the medical system. She may continue to seek out an expert who will believe her, or she might well isolate herself from those who question her condition.

One of the strange things about our medical system is that any physician can prescribe any medication. A psychiatrist can write a prescription for fluconazole to treat athlete's foot, and a dermatologist can write a prescription for pimozide to treat delusory parasitosis. Perhaps pragmatism trumps principle when it comes to the responsibilities of physicians. Maybe it is better that the dermatologist makes a psychological diagnosis and provides an antipsychotic than that the patient be allowed to suffer because the illness involves the brain rather than the skin. It's a tough call, but dermatologists have developed guidelines stating that they should play the role of psychologists only when a patient is hostile to the possibility of a mental disorder and is not depressed, psychotic, or in need of psychotherapy.[68]

All of this assumes that the patient is not actually infested, and, as with illusory parasitosis, being sure that there are no actual arthropods can be a challenge. As one expert noted, "Despite the presence of personality traits or psychiatric symptoms which corroborate a diagnosis of delusions of parasitosis, the patient is sometimes infested."[69] This is analogous to the observation that "just because you're paranoid doesn't mean they aren't after you."

Consider the case of two elderly ladies who complained to the Los Angeles County Health Department that some "bug" was biting them.[70] The ladies sent in bits of cloth, paper, and tape supposedly containing the culprits—a classic indicator of delusory parasitosis and a slam dunk *folie à deux*. Nobody could find an infestation, case closed—or at least referred to the mental health folks. Except that in the fourth delivery of putative specimens from the ladies, a technician discovered northern fowl mites (one must admire the health department's willingness to continue checking for actual arthropods). These parasites strongly favor birds but will deign to bite humans. When an exterminator removed birds' nests from the elderly ladies' house and sprayed the residence, the problem was solved.

Just as entomologists shouldn't make psychiatric diagnoses, physicians are not generally qualified to make entomological diagnoses. Even though physicians might believe that "a rudimentary understanding of arachnid life cycles, biology and behavior, if not already known, is easily obtained by any therapist by consulting medical entomology texts or local entomologists,"[71] the resultant knowledge may be just enough to make a doctor dangerous. After all, we wouldn't want an entomologist thumbing through a dermatology manual and making diagnoses by matching pictures to a patient's condition.

In this regard, consider the story of a woman who bounced among various physicians seeking relief for her mite infestation.[72] She'd show up with jars and folded tissues supposedly containing the vermin to no avail, until a psychiatrist finally told her that he found something that looked like a headless mite, thereby affirming her delusion. Fortunately, a competent exterminator came to her house, examined her "specimens," and showed her the difference between the dust that she had collected and genuine mites. After some hours, her convictions weakened, and she began to accept that her mites were not real. The woman agreed to see a doctor (presumably one who specialized in mental illness rather than headless arthropods), and at the time of the report she had been free of her delusions for four years. But this resolution was easy compared to the classic case of Jay Traver.

For decades, whenever scientists or doctors admonished others to be careful before making a diagnosis of delusory parasitosis, they referred to a 1951 paper published in the *Proceedings of the Entomological Society of Washington*, a reputable journal largely focused on taxonomy.[73] The author, Jay Traver, was a zoologist at the University of Massachusetts who described her infestation by a rare mite after having been rebuffed by doctors for years.

Traver's ordeal began in 1934 with a persistently itchy scalp that by 1943 she interpreted as "sensations as of some arthropod crawling, scratching and biting."[74] By 1951, she believed that the mites had also infested her ears, eyebrows, lips, nostrils, throat, trachea, bronchi, neck, back, abdomen, chest, shoulders, arms, wrists, and knees, as well as the areas between her toes and her fingers. The poor woman wrote that "mites could be felt running about below the thickened epidermal masses, and causing their host almost to tear off bits of the scalp."[75] She developed secondary infections from trying to excavate the creatures from the "burrows" they had tunneled beneath her flesh.

Traver was desperate for relief and treated herself with a staggering array of chemicals, including soap containing 1 percent mercuric iodide, sulfur and zinc oxide ointments, Lysol (at four times the concentration recommended for disinfecting surfaces), hot boric acid, carbolic acid, sulfathiazole (a rather toxic antimicrobial sulfa drug), DDT powder, lindane (an organochlorine insecticide), kerosene emulsion with vinegar, various alcohols, a paste of sulfur and Crisco, benzyl benzoate (a topical treatment for scabies, at twice the recommended rate), Sergeant's Mange Cure (for pets), and perhaps most remarkably a 40 percent solution of sodium hypochlorite (eight times more concentrated than commercial bleach) followed by 5 percent hydrochloric acid.

After dermatologists and neurologists concluded that her infestation was imaginary, she redoubled her efforts to prove the existence of parasites. At last, she was able to extract samples of mites: "This was a tedious process, and the number of mites actually captured is surprisingly low. This does not mean, to me at least, that there were not many more mites present."[76] She identified the creatures as *Dermatophagoides scheremelewskyi*. This rare species was discovered by a Russian acarologist in 1864, who found the mite in rodent and bird nests as well as such unlikely sources as human urine and sputum samples.[77] Traver thought her infestation had come from a neighbor's dog or cat.

So with the mite identified, the symptoms explained, and the study published, the case was closed. Except for one thing: in 1993 it was revealed that Traver's paper was "one of the most remarkable mistakes ever published in a scientific entomological journal."[78]

Alex Fain, an internationally renowned acarologist, examined Traver's specimens and determined that they were the common house dust mite. The ubiquity of dust mites in our homes was not known until the mid-1960s, when Fain published his research on this subject. This explained the Russian scientist's "discovery"—the dusty conditions of his laboratory supported a thriving population of mites that he misidentified after they wandered into his various specimen containers. Dust mites don't burrow into human tissue, so Traver's account was a vivid, poignant, and tragic autobiography of a woman driven to desperate measures to affirm the reality of her delusion.

One might think that the case of Jay Traver would have been a lesson that would put editors of scientific journals on alert. However, a remarkable reprise played out with an even more bizarre paper in the *Journal of the New York Entomological Society*—a publication with a century-long history of credible

Figure 6.3
Common dust mites, such as those shown here, were the only creatures that Jay Traver actually found in her classic 1951 report of a supposedly new mite capable of infesting humans. Although Traver believed that she was infested with a rare species, her mistake was revealed in 1993, and her scientific paper turned out to be a tragic account of delusory parasitosis (image by Gilles San Martin through Creative Commons).

science.[79] In 2004, Deborah Altschuler and her colleagues described their attempt to discover an underlying cause for twenty cases that had been diagnosed as delusory parasitosis. For six months they looked at microscope slides of skin scrapings and found nothing.

Then they came across a few strange reports in obscure outlets of Collembola being found on humans. Collembola, or springtails, are tiny insects that live in moist habitats and feed on detritus. There was one credible, peer-reviewed paper describing a family in Texas infested with springtails, but in that case there was no evidence of irritation or inflammation. Nonetheless, Altschuler and her team had their suspect.

They went back to their microscopes and stared until they finally found the culprit. Using a spectacularly flawed, uncontrolled, and biased protocol that virtually assured the discovery of some "common denominator," the investigators eventually managed to identify the remains of Collembola in eighteen of the subjects. Their published evidence consisted of three photomicrographs.

According to the authors, "Identification of Collembola in scrapings from symptomatic study participants required intensive scrutiny of the photographs and was initially very difficult."[80] Most likely because nothing was there. Looking at the images is much like staring at clouds and seeing sheep—or springtails. The charitable explanation for the authors' claim is that they fell prey to pereidolia.[81] This is a type of illusion in which a vague or obscure stimulus is perceived as something clear and distinct, as when an amorphous blob of tissue viewed at just the right angle and with just the right lighting can look like the outline of a Collembola.

Just as we have to point out to others how a cloud looks like a cavorting lamb, Altschuler's team marked the outline of a springtail in one of the figures for the reader. In the other image, they labeled the body parts of the purported insect. The authors admitted: "It was only after intense scrutiny that they were recognized. In addition, because the Collembola were not always intact or completely in focus, they were difficult to discern."[82]

Scientists at the Bohart Museum of Entomology, one of the top entomology departments in the United States, summarily dismiss the conclusions drawn by Altschuler and her colleagues with an unusually direct rebuke (scientists generally favor passive aggression):

> Collembola lack the structures necessary to enter human skin, either to feed or to lay eggs [these insects are termed "entognathus," meaning that their mandibles are held deep within their heads]. The authors of the single publication proposing to have found Collembola in human skin (2004. New York Entomological Society) manipulated photographs in Adobe Photoshop to supposedly show the presence of Collembola and their eggs. The quality of the photographs and the quality of objects in the photographs was so poor that it is quite possible to see almost any kind of animal in them. The journal should be embarrassed for having published this work.[83]

The US Centers for Disease Control and Prevention (CDC) concurs with the implausibility that 90 percent of people suffering from delusory parasitosis are actually infested with springtails.[84] And most recently, a categorical refutation of the Altschuler paper was published by a pair of Collembola specialists:

> None of the images presented in Altschuler et al. (2004), except possibly the "enhanced" version of Figure 2 (p. 91), bear the slightest resemblance to any springtail or springtail body parts, nor can the pieces in Figs. 1 and 3 be reasonably construed to represent a part of any particular organism. . . . Under the assumption that the best images were chosen for the article, the assertion that Collembola are the material cause of "delusory parasitosis" is not supported by the evidence. . . . Apparently, no Collembola expert was consulted before publication, and the more than 300 slides of scrapings made in this study have never been made available to a Collembola expert for examination of the supposed specimens.[85]

Nevertheless, the National Pediculosis Association (pediculosis being an infestation by lice and Altschuler being the president of the NPA) continues to push the case for Collembola infesting humans. In a bizarre echoing of the unwavering insistence of a victim of delusory parasitosis, the organization stridently insists that the scientific world is mistaken, that its beliefs are veridical, and that insects are present even if nobody else can see them.[86]

The ramifications of bad entomological science extend into other disorders. At least some of those who suffer with Morgellons disease are displeased with Altschuler's shenanigans. In the early days, this disease was often diagnosed as delusory parasitosis. However, it soon became evident that only some of the symptoms resemble those reported by people who imagine they are infested with insects. The CDC describes the symptoms as including crawling, biting, and stinging sensations, but these are accompanied by the presence of visible granules, threads, fibers, or black speck-like materials on or beneath the skin, rashes, sores, and various nondermatological problems such as fatigue, confusion, short-term memory loss, joint pain, and changes in vision.[87]

But just as the CDC was beginning to take Morgellons seriously, along came Altschuler's publication declaring that delusory parasitosis is really due to Collembola. This preposterous research did not help the credibility of those suffering from dermatological conditions of undetermined origin. However, the CDC has undertaken an epidemiological investigation, so the damage has been contained—at least within the medical community.

Given that the cause of Morgellons remains unexplained, those with the condition are understandably desperate for answers. And entomological confusion abounds. Postings on the Morgellons Disease Research online forum have included statements such as this one:

> I definitely had these [Collembola] on my body at some point (got photos!), and showed this article to my doctor last year. . . . they may like to live under the skin at the root of hair follicles.[88]

What Morgellons will turn out to be is anyone's guess. The temptation to make the relatively easy diagnosis of delusory parasitosis is understandable. However, just as people with psychiatric disorders who feel itchy might really be infested, patients who experience crawling, tingling sensations might have a dermatological disorder. In either case, a mistaken diagnosis of delusory parasitosis leaves the patient suffering. And in any case, a search for Collembola is almost sure to be futile.

UNRAVELING THE INFESTED MIND

We started out hoping to cut through the Gordian knot of false perceptions and beliefs. Psychologists have pulled loose and defined the threads of hallucination, illusion, and delusion, and these isolated phenomena seem clear enough. However, it seems that the inner world is often too snarled for any one thread to explain our experiences.

In my case, the lucid dream (a kind of hallucination) of being swallowed by an indifferent organic mass came rushing back amid the pandemonium of grasshoppers. For hours I was plagued by the tactile illusion that insects were on me and in my clothes, when there was nothing but grass seeds and drops of sweat. And now, as I try to recount that day accurately, I wonder if my recollections are woven into a confabulation—the less pathological cousin of the delusion, in which one forms a coherent but unintentionally mistaken belief about past events through distorted memory.[89]

Perhaps the scariest thing about hallucinations, illusions, and delusions is the sense that we are losing our minds. Modern psychology does not define insanity—the *Diagnostic and Statistical Manual of Mental Disorders* has no listing for this condition. However, the legal definition resonates with our intuition: "mental illness of such a severe nature that a person cannot distinguish fantasy from reality" (the definition goes on for another 362 words, making it one of the lengthier entries—more than four times longer than the definition for "pain and suffering" and twice as long as that for "justice").[90]

We want our internal world to correspond to the actual world, and even fleeting disconnections can leave us profoundly perplexed and emotionally shaken. If you managed to finish this chapter without an urge to scratch, then maybe you're particularly resistant to imaginary sensations, perceptions, and beliefs. Otherwise, perhaps you've latched onto one of those tangled threads that lead to the infested mind.

NOTES

1. Jan Dirk Blom, *A Dictionary of Hallucinations* (New York: Springer, 2009).
2. Ibid.; Michael James Burns, "Delirium tremens (DTs)," Medscape, June 29, 2011, http://emedicine.medscape.com/article/166032-overview (accessed October 31, 2011); John P. Cunha, "Alcoholism," eMedicineHealth, http://www.emedicinehealth.com/alcoholism/article_em.htm (accessed October 31, 2011).
3. Blom, *A Dictionary of Hallucinations*, 103.
4. Ibid.
5. Wolfgang W. May and Margaret S. Terpenning, "Delusional parasitosis in geriatric patients," *Psychosomatics* 32 (1991): 88–94.
6. H. G. Scott and J. M. Clinton, "An investigation of 'cable mite' dermatitis," *Annals of Allergy* 25 (1967): 409–15.
7. W. J. Simpson, "Cable bugs—Mysterious biting insects or faulty diagnosis?," *Parasitology Today* 3 (1987): 323–24.
8. Ibid.
9. Phillip Weinstein, "Insects in psychiatry," *Cultural Entomology Digest*, no. 2 (1994), Insects.org, http://www.insects.org/ced2/insects_psych.html (accessed October 31, 2011).

10. Dario Maestripieri, "Vigilance costs of allogrooming in macaque mothers," *American Naturalist* 141 (1993): 744–53.
11. Isabelle Dean and Michael T. Siva-Jothy, "Human fine body hair enhances ectoparasite detection," *Biology Letters* 8 (2012): 358–61.
12. Atul Gawande, "The itch," *New Yorker*, June 30, 2008), http://www.newyorker.com/reporting/2008/06/30/080630fa_fact_gawande (accessed October 31, 2011)
13. Ibid.
14. Weinstein, "Insects in psychiatry."
15. Ibid.
16. Simpson, "Cable bugs."
17. May R. Berenbaum, *Bugs in the System: Insects and Their Impact on Human Affairs* (New York: Helix Books, 1995).
18. William G. Waldron. "The entomologist and illusions of parasitosis," *California Medicine* 117 (1972): 76–78.
19. Simpson, "Cable bugs."
20. Waldron, "The entomologist and illusions of parasitosis."
21. Ibid.
22. Meredith Etherington-Smith, *The Persistence of Memory: A Biography of Dalí* (New York: Random House, 1992); Salvador Dalí, *The Secret Life of Salvador Dalí* (1942; repr., Whitefish, MT: Kessinger, 2010).
23. Luis Buñuel, *My Last Sigh* (New York: Vintage, 1982), quoted in Berenbaum, *Bugs in the System*, 304.
24. Dalí, *Secret Life.*
25. Lloyd A. Miller, "An account of insect hallucinations affecting an elderly couple," *Canadian Entomologist* 86 (1954): 455–57.
26. Ibid., 456.
27. G. E. Berrios, "Delusional parasitosis and physical disease," *Comprehensive Psychiatry* 26 (1985): 395–403; Weinstein, "Insects in psychiatry."
28. Weinstein, "Insects in psychiatry"; Karl Axel Ekbom, "Der praesenile Dermatozoenwahn," *Acta Psychiatrica et Neurologica Scandinavica* 13 (1938): 227–59.
29. Waldron, "The entomologist and illusions of parasitosis."
30. Berrios, "Delusional parasitosis and physical disease."
31. W. R. Lee, "Matchbox sign," *Lancet* 322 (1983): 457–58.
32. Weinstein, "Insects in psychiatry"; Albert H. Schrut and William G. Waldron, "Psychiatric and entomological aspects of delusory parasitosis: Entomophobia, acarophobia, dermatophobia," *Journal of the American Medical Association* 186 (1963): 429–30.
33. Waldron, "The entomologist and illusions of parasitosis."
34. Weinstein, "Insects in psychiatry."
35. N. J. Hunt and V. R. Blacker, "Delusional parasitosis," *British Journal of Psychiatry* 150 (1987): 713–14.
36. Ibid., 713.
37. M. L. Bourgeois, P. Duhamel, and H. Verdoux, "Delusional parasitosis: Folie à deux and attempted murder of a family doctor," *British Journal of Psychiatry* 161 (1992): 709–11.
38. Ibid.
39. Laurie Lawlor, *This Tender Place: The Story of a Wetland Year* (Madison: University of Wisconsin Press, 2007), 94; Merrill Denison, *Klondike Mike: An Alaskan Odyssey* (Whitefish, MT: Kessinger, 2005), 209.
40. Charles St. Charles, "Field notes: The ends of the day," Nature of the Wild, http://natureofthewild.com/FieldNotes/TheEndsOfTheDay.html (accessed November 19, 2012).

41. J. Kenneth Grace and David L. Wood, "Delusory cleptoparasitosis: Delusions of arthropod infestation in the home," *Pan-Pacific Entomologist* 63 (1987): 1–4.
42. Ibid.
43. Michael A. Marschall, Rudolph F. Dolezal, Mimis Cohen, and Stephanie F. Marschall, "Chronic wounds and delusions of parasitosis in the drug abuser," *Plastic and Reconstructive Surgery* 88 (1991): 328–30.
44. Waldron, "The entomologist and illusions of parasitosis."
45. Gawande, "The itch."
46. Schrut and Waldron, "Psychiatric and entomological aspects of delusory parasitosis," 430. See also J. Mumford, "Entomophobia: The fear of arthropods," *Antenna* 6 (1982): 455–57.
47. Schrut and Waldron, "Psychiatric and entomological aspects of delusory parasitosis."
48. Ibid.
49. Shivani Chopra, "Delusional disorder," Medscape, March 10, 2011, http://emedicine.medscape.com/article/292991-overview#a1 (accessed October 31, 2011); Marschall et al., "Chronic wounds and delusions of parasitosis in the drug abuser"; Berenbaum, *Bugs in the System*; Weinstein, "Insects in psychiatry."
50. May and Terpenning, "Delusional parasitosis in geriatric patients"; Weinstein, "Insects in psychiatry"; Bourgeois et al., "Delusional parasitosis."
51. Alan Lyell, "Delusions of parasitosis," *British Journal of Dermatology* 108 (1983): 485–99.
52. Chopra, "Delusional disorder."
53. Ibid.
54. Weinstein, "Insects in psychiatry."
55. Author interview with Douglas Smith, September 16, 2011.
56. J. H. Poorbaugh, "Cryptic arthropod infestations: Separating fact from fiction," *Bulletin of the Society for Vector Ecology* 189 (1993): 3–5.
57. C. Pomerantz, "Arthropods and psychic disturbances," *Bulletin of the Entomological Society of America* 5 (1959): 65–67; Miller, "An account of insect hallucinations affecting an elderly couple."
58. Pomerantz, "Arthropods and psychic disturbances."
59. Ibid.
60. May and Terpenning, "Delusional parasitosis in geriatric patients," 93.
61. Berrios, "Delusional parasitosis and physical disease."
62. Weinstein, "Insects in psychiatry."
63. Berrios, "Delusional parasitosis and physical disease."
64. Chopra, "Delusional disorder"; Aaron T. Beck, *Cognitive Therapy and the Emotional Disorders* (New York: Penguin, 1991).
65. May and Terpenning, "Delusional parasitosis in geriatric patients."
66. L. E. Smith, "Analysis of commercial phenothiazine used as an insecticide," *Industrial and Engineering Chemistry, Analytical Edition* 10 (1938): 60; M. L. Colglazier, A. O. Foster, F. D. Enzie, and D. E. Thompson, "The anthelmintic action of phenothiazine and piperazine against *Heterakis gallinae* and *Ascaridia galli* in chickens," *Journal of Parasitology* 46 (1960): 267–70.
67. Weinstein, "Insects in psychiatry."
68. Maximilian E. Obermayer, "Dynamics and management of self-induced eruptions," *California Medicine* 94 (1961): 61–65.
69. Weinstein, "Insects in psychiatry."
70. William G. Waldron, "The role of the entomologist in delusory parasitosis (entomophobia)," *Bulletin of the Entomological Society of America* 8 (1962): 81–83.

71. Weinstein, "Insects in psychiatry."
72. Pomerantz, "Arthropods and psychic disturbances."
73. Jay R. Traver, "Unusual scalp dermatitis in humans caused by the mite, *Dermatophagoides*," *Proceedings of the Entomological Society of Washington* 53 (1951): 1–25.
74. Ibid., 1.
75. Ibid., 4.
76. Ibid., 10.
77. Poorbaugh, "Cryptic arthropod infestations."
78. Ibid., 3.
79. Deborah Z. Altschuler, Michael Crutcher, Neculai Dulceanu, Beth A. Cervantes, Cristina Terinte, and Louis N. Sorkin, "Collembola (springtails) (Arthropoda: Hexapoda: Entognatha) found in scrapings from individuals diagnosed with delusory parasitosis," *Journal of the New York Entomological Society* 112 (2004): 87–95.
80. Ibid., 89.
81. May Berenbaum, "Face time," *American Entomologist* 51 (2005): 68–69; C. S. H. Lim, S. L. Lim, F. T. Chew, T. C. Ong, and L. Deharveng, "Collembola are unlikely to cause human dermatitis," *Journal of Insect Science* 9, no. 3 (2009): 1–5; "Pareidolia," The Skeptic's Dictionary, http://skepdic.com/pareidol.html (accessed October 31, 2011).
82. Altschuler et al., "Collembola (springtails)," 91.
83. Bohart Museum of Entomology, "Human skin parasites and delusional parasitosis," University of California, Davis, Department of Entomology, http://delusion.ucdavis.edu/otherparasites.html (accessed October 31, 2011).
84. "An Emerging Health Crisis—Where's Dr. House When You Need Him?," National Pediculosis Association, press release, May 19, 2005, http://www.headlice.org/news/2005/0519.htm (accessed April 1, 2013).
85. Kenneth A. Christiansen and Ernest C. Bernard, "Critique of the article 'Collembola (springtails) (Arthropoda: Hexapoda: Entognatha) found in scrapings from individuals diagnosed with delusory parasitosis,'" *Entomological News* 119 (2009): 538.
86. "Skin scraping/collembolan research," National Pediculosis Association, http://www.headlice.org/report/research/index.html (accessed October 31, 2011).
87. "CDC Study of an unexplained dermopathy," Centers for Disease Control and Prevention, http://www.cdc.gov/unexplaineddermopathy (accessed April 1, 2013); "Report of the External Peer Review of CCID's Unexplained Dermopathy (UD) Project, 2009", John Bennett (redactor), http://www.cdc.gov/unexplaineddermopathy/docs/external_peer_review.pdf (accessed April 17, 2013).
88. "Message Board," Morgellons Disease Research, http://www.morgellons-disease-research.com/Morgellons-Message-Board/index.php (accessed October 31, 2011).
89. Robyn Langdon and Tim Bayne, "Delusion and confabulation: Mistakes of perceiving, remembering and believing," *Cognitive Neuropsychiatry* 15 (2010): 319–45.
90. American Psychiatric Association, *Diagnostic and Statistical Manual of Mental Disorders*, 4th ed., text revision (Arlington, VA: American Psychiatric Publishing, 2000); "Insanity," Law.com, http://dictionary.law.com/Default.aspx?selected=979 (accessed November 13, 2011).

CHAPTER 7

Treating the Infested Mind: Exterminating Entomophobia

At least 6 percent of Americans suffer from entomophobia, and some studies report that as many as one in five women experience this condition in any given year.[1] Conservatively, twenty million people live with irrational fears of insects and their kin. From a medical perspective, most somatic and mental disorders cause human suffering either because we are unable to diagnose the condition readily or because we lack the ability to treat the illness. Chronic fatigue syndrome is an instance of the former, lupus is a case of the latter, and pancreatic cancer is an example of both. But entomophobia does not fall into either category. While harmless multilegged creatures are likely to bear the insecticidal brunt of our debilitating fear, the real tragedy is that the suffering—both the convulsions of the poisoned insects along with other nearby creatures and the terrible strain on the afflicted humans along with their family members and friends—is unnecessary.

ENTOMOPHOBIA: CHOOSING TO SUFFER

Entomophobia is readily diagnosable. An interview method developed by David Barlow, founder of the Center for Anxiety and Related Disorders at Boston University, generates a remarkable 86 percent consensus among therapists on the diagnosis of specific phobias—a higher rate of agreement than for any other anxiety or mood disorder.[2] Psychologists might argue with one another about whether a patient is suffering depression or bipolar disorder, but there's not much debate about someone with a specific phobia. There may be some confounding with disgust, but in any case the result is a consistent, unreasonable, and marked aversion. When there is disagreement, the conflict most often arises either because practitioners differ as to whether the symptoms are sufficiently severe to qualify for a clinical condition or because the patient has provided different information to various therapists.[3] But in medicine, it is often the case that diagnostic methods are more readily available than dependable treatments.

Specific phobias, of which entomophobia is a classic example, are among the most treatable of all psychological disorders. In part, this may be a function of low comorbidity, meaning that only 15 percent of patients with a particular phobia (say, a fear of spiders) exhibit additional specific phobias (for example, fear of heights, closed spaces, or clowns).[4] A somewhat higher proportion meet the criteria for a mood or anxiety disorder, but for the most part people with a persistent, irrational, and excessive fear of insects have the best prognosis of anyone with a mental illness. A remarkable 90 percent of patients achieve clinically significant, long-lasting improvement with as little as one treatment session.[5]

So we have a medical paradox—despite solid diagnostic methods and effective treatments, only one out of eight people with specific phobias seeks help. It's not as if these folks aren't miserable, given that the most common symptoms include a racing heart, shortness of breath, muscle tightness, trembling, and feelings of impending doom.[6] Rather, it appears that most specific phobias are easier to dodge in daily life than are other mood or anxiety disorders. The fear is

Figure 7.1
Women experience anxieties and phobias more often than do men. Whereas anxieties are notoriously difficult to resolve, specific phobias (such as entomophobia) are the most treatable psychological disorders, with 90 percent of patients showing marked improvement. Unfortunately, few people seek treatment because most choose to avoid insects or other triggers of irrational fear (image from TWINTHOMAS through Creative Commons).

explicitly linked to a stimulus that people can often avoid without great disruption to their lives. In short, phobics develop "workarounds" so that they don't experience the debilitating symptoms of acute and severe fear. Entomophobes simply don't go into the storage shed, look under the sink, or move to Florida.

But what if you're tired of avoiding sheds, sinks, and sun?

THE THERAPIST'S TOOLBOX

Fear—along with disgust—is a complicated, aversive phenomenon that gives rise to a mix of responses (emotional, cognitive, physiological, behavioral) triggered by a variable set of stimuli (an entomophobe might fear bees and moths but not cockroaches or spiders) and arising from tangled origins (evolutionary, familial, and sociological). So is it any wonder that therapies are also complicated? Although many therapists use a combination of approaches, to make sense of these tactics I'll parse them into operational categories beginning with those of dubious efficacy.

Pop a Pill

Name: Salvador Dalí **Date: May 11, 1929**
Address: 11 rue Poulbot, Paris **Age/Wt: 25/64 kg**
Rx Alprazolam 1 mg
Disp #14
Sig: Take 1 tablet bid x 7 days

Of course, Salvador Dalí (our prototypical entomophobe) did not have access to modern psychoactive drugs such as alprazolam, which decreases abnormal brain activity. Today's physicians have a pharmacopeia of options for treating anxiety disorders (tranquilizers, anxiolytics, beta blockers, etc.). However, the general sense among mental health professionals is that medications offer little benefit for specific phobias, even when combined with other forms of therapy.[7] Prescribing drugs might be justifiable when the phobia is so severe that it poses an immediate threat to the individual, but in the vast majority of cases, changing behavior rather than masking symptoms is the way to go.

On the Couch

"So, Mr. Dalí," the psychoanalyst began, "you were afraid of your father and terrified by grasshoppers." The artist nodded his agreement. "And now you've painted a picture that you call *The Great Masturbator* that includes a swarm of ants on the abdomen of a grasshopper that is itself feeding on a distorted human face below the image of a woman whose mouth is approaching a thinly clad male crotch." Dalí shrugged his consent. "It seems we have some work ahead of us," the therapist mused, sitting back in her chair and chewing on her pencil, which was, in this context, nothing but a pencil.

Although the dialogue is fictional, Dalí found inspiration in Sigmund Freud's work—and one might presume that a contemporary psychoanalyst would have an extraordinarily interesting time working with Dalí to unearth the disgust and confront the fear that fueled his repressed desires. Despite contentions that "for chronic phobias the prognosis is good if the underlying conflict can be dealt with,"[8] few data are available concerning the effectiveness of psychotherapy in treating entomophobia.[9]

Relax . . .

"Close your eyes and let yourself relax, Mr. Dalí." The artist began to breathe deeply. The hypnotherapist continued her soothing banter: "The deeper you relax, the better you feel. You are drifting, floating." As the artist's head nodded forward, he slid into a state of focused consciousness. In this trance, he was highly amenable to ideas that accorded with his desires, such as the suggestion that he would feel tranquility in the presence of grasshoppers.

If Dalí were alive today, he might be reassured by those who promise to recondition his fear through hypnosis and its therapeutic neighbor, neurolinguistic programming:

> A seventeen year old girl . . . unable to say the word "cockroach" out loud . . . had learned her fear [when] hundreds of the insects swarmed over [her] feet. . . . I massively disassociated her from the experience . . . and attached the "humour" feelings to thoughts and memories of the insects. It took 20 minutes . . . and I was almost as amazed as she was that she was totally OK as a result.[10]

If this anecdote seems too good to be true, it probably is. There isn't much clinical evidence that hypnotic reframing is effective, and even less evidence supports neurolinguistic programming.[11]

Would You Look at That Thing!?

"Mr. Dalí, imagine that you're in a darkened room. Can you picture this in your mind?" The artist gave a cautious nod. "As the lights come up, you realize that the walls are bristling with grasshoppers." Dalí began to tremble. She reminded him, "Stay with the image and don't forget what you practiced with your breathing." As he covered his face, the therapist continued: "The grasshoppers begin leaping and they land on your clothing and in your hair. You can feel their spiny legs clinging to you." Dalí shuddered as the therapist pushed on.

Exposure therapy entails putting the patient in imaginative or physical proximity—even direct contact—with the object of fear. This approach takes a variety of forms, from the harsh to the gentle. Trauma therapists maintain that the phobic individual has developed a dual-belief system that allows fear to persist.[12] When the spider is present, the arachnophobe believes that disaster is imminent. But in the absence of the creature, the individual understands that spiders are relatively harmless. So the beliefs are contradictory and conditional. As the arachnophobe approaches the spider, rational thought gives way to terrifying fantasy supported by cognitive distortions, visual images, and bodily sensations. So what better way to defeat the false beliefs than by showing them to be absurd—suddenly, dramatically, and decisively?

In implosive therapy, the patient is thrust into an imaginary scenario that forces a confrontation with the person's catastrophic fantasies.[13] An arachnophobic individual might envision being caught in the web of a giant spider in a scene out of the *The Fly* (the one where the tiny human-headed fly cries, "Help me!" as the ravenous spider slowly approaches its meal).

A variation on this theme emerged in the 1990s: eye-movement desensitization and reprocessing. The technique involves the phobic patient imagining an encounter with the feared stimulus while tracking the therapist's finger as it moves side to side. The notion is that, "through the eye movements, negative memories lose their pathogenic character and will be assimilated." If that sounds weird, it is—and at least for arachnophobes it doesn't work or, at most, adds little to the basic treatment.[14]

Moving your eyes back and forth might not accomplish much, but there is another way to confront negative thoughts—by being placed in actual contact with the feared object in a process called flooding.[15] A therapist might have taken Dalí to a field crawling with grasshoppers and had him walk through the grass as the insects leapt and ricocheted off his body, thereby forcing the artist to face his terror—and to exhaust his irrational fear in a supportive context. A draw in Whalen Canyon, Wyoming, might have been just the place. But flooding can, when administered by those who seek to extract information rather than provide solace, become a kind of drowning—an entomophobic waterboarding of sorts.

In the hands of a skilled torturer, sudden and forced contact with insects can become a singular source of agony for a prisoner with a deep-seated fear of insects. Documents released by President Barack Obama in April 2009 revealed that the CIA had intended to use insects as part of an Orwellian strategy (remember Winston and the rats in *1984*?) to extract information from Abu Zubaydah, an alleged high-ranking member of al-Qaeda.[16] A 2002 memo from the US Justice Department ruled that exploiting the captive's entomophobia did not constitute torture based on the absurd contention that a "reasonable" person would not fear bodily harm (the whole point of the CIA plan was to tap into an *unreasonable*, but profoundly powerful, psychological vulnerability of the prisoner):

> In addition to using the containment boxes alone, you also would like to introduce an insect into one of the boxes with Zubaydah. As we understand it, you plan to inform Zubaydah that you are going to place a stinging insect into the box, but you will actually place a harmless insect in the box, such as a caterpillar. If you do so, to ensure that you are outside the predicate act requirement, you must inform him that the insects will not have a sting that would produce death or severe pain. If, however, you were to place the insect in the box without informing him that you are doing so, then, in order to not commit a predicate act, you should not affirmatively lead him to believe that any insect is present which has a sting which could produce severe pain or suffering or even cause his death [section of text redacted] so long as you take either of the approaches we have described, the insect's placement in the box would not constitute a threat of severe physical pain or suffering to a reasonable person in his position. An individual placed in a box, even an individual with a fear of insects, would not reasonably feel threatened with severe physical pain or suffering if a caterpillar was placed in the box. Further, you have informed us that you are not aware that Zubaydah has any allergies to insects, and you have not informed us of any other factors that would cause a reasonable person in that same situation to believe that an unknown insect would cause him severe physical pain or death. Thus, we conclude that the placement of the insects in the confinement box with Zubaydah would not constitute a predicate act.[17]

The Justice Department's oxymoronic legal description of a "reasonable" entomophobe being forced into a box with an insect and its ridiculously contrived psychological assessment that entomophobia entails the belief that insects will inflict severe pain constitute either appalling scientific ignorance or perversely twisted reasoning. Fortunately for phobic individuals, there are far fewer torturers than therapists. And as for the latter, there's no compelling evidence that traumatic flooding works better than kinder and gentler versions of exposure therapy.

* * *

"Mr. Dalí, imagine that there's a grasshopper in a jar on the other side of the room." He gave a slight shudder. "Next, I'm asking that you just peek at it." As the artist clenched his fists and shuddered, the therapist instructed him to use the relaxation techniques that he had practiced. As he did so, she went to the next step: "Very good. Next I want you to imagine edging closer to the jar until it is an arm's length away." When his trembling resumed, she again coaxed him into calmness. The session continued until she asked him, "Now, what if the grasshopper leaps out of the jar?"

In systematic desensitization, a therapist takes the phobic patient through a series of increasingly terrifying mental images, allowing the individual to relax after each step until the waves of panic subside.[18] This image-based approach was developed by Joseph Wolpe fifty years ago. He figured that once the patient's distorted view of reality was fully confronted and recognized, it could be transformed. Various techniques emerged for modifying mental images, such as time projection (e.g., imagining a future in which grasshoppers are no longer frightening), substitution of positive imagery (e.g., imagining that grasshoppers are gentle little friends), the use of coping models (e.g., imagining that one is a world-renowned entomologist supremely adept at handling grasshoppers), and de-catastrophizing (e.g., imagining that the grasshopper escapes into the room and realizing that nothing awful happens, which is like cognitive flooding without the trauma).[19]

Today, we can substitute technology for imagination—or at least that was the plan. Psychologists have tried using computer images of spiders to desensitize patients. Consider the meticulously documented case of a ten-year-old girl who had discovered a large spider in her room a few years earlier.[20] The child had been frozen with terror for several hours until her father came home and removed the creature. It probably didn't help that the girl's mother had been treated for arachnophobia and provided a vivid model for a debilitating fear of spiders. The therapist used a "hierarchically structured confrontation with spiders presented on a computer screen" in an effort to incrementally induce the child into realizing that actual spiders aren't horrifying. But in the end this approach to desensitization didn't work, as the computer images lacked the vivid intensity needed to initiate the corrective process. More recent efforts to use interactive animations to model exposure have also failed,[21] and there is no reason to believe that the iPhone app for arachnophobes is worth the $2.99 download price. Until technology becomes more psychologically convincing, there's one thing better than virtual reality: real reality.

Wolpe's use of imaginary encounters has generally fallen out of favor among experts. Many psychologists maintain that exposure to the feared stimulus itself is necessary, and often sufficient, to treat phobic individuals.[22] And in vivo ("in life") therapy is reportedly quite effective in treating spider phobias. Research suggests that a session with the real deal requires less time, particularly if the therapist doesn't bother with periods of relaxation, which don't appear to improve results.[23]

So a contemporary therapist using systematic desensitization might have put poor old Dalí in a room with a real grasshopper while gently and firmly urging him toward making contact with the living, breathing creature. Just such an approach is featured on the Animal Planet program *My Extreme Animal*

Phobia—and Dalí might even have been chosen to appear on this strange reality show. Animal Planet's website promotes the show with this description:

> It may seem completely reasonable to be deathly afraid of snakes, but what if you were terrified of moths, grasshoppers or even puppies? A phobia can become so intense it transforms everyday behavior in a way that could make you laugh—if it didn't actually reduce you to terror or tears![24]

After just two to three hours of in vivo exposure therapy with systematic desensitization, about 90 percent of people with zoophobias show marked improvement. And follow-up studies reveal that the recovery lasts at least four years.[25] Rather surprisingly, a study of people suffering from a fear of snakes found that those who experienced only a partial reduction of their fear during treatment were less likely to have their fear return than were those who reported a complete elimination of fear.[26] So there might be good news even for those who continue to feel some trepidation after therapy. But there's some bad news for those who feel disgusted.

The need to put an entomophobe in contact with an actual insect would have been particularly important in the case of Dalí. It appears that the artist responded to grasshoppers with a mix of fear and disgust (recall his encounter with the "slobberer" fish, which he equated with a grasshopper). When the latter emotion is involved, images and pictures are unlikely to be effective, and even with in vivo exposure therapy the prognosis is guarded.[27]

Arachnophobes who are also highly sensitive to disgust show less improvement with exposure therapy than do individuals with normal disgust sensitivity. Moreover, disgust itself is much more resistant to treatment than is fear. Studies of arachnophobes have found that when exposure therapy was successful in terms of fear, the patients often retained their disgust. Some researchers propose that better results may ensue if therapists employ longer or more frequent exposures to achieve disgust habituation, while others suggest that therapists expose patients to explicitly pleasant stimuli (e.g., eating tasty food or listening to favorite music) during or after spider encounters.[28]

While disgust requires innovative modifications of therapy, such tweaking of practices is familiar to psychologists, who continue to fine-tune their methods for treating straightforwardly simple phobias in the search for ever greater efficiency and efficacy.

* * *

In terms of efficiency and hence cost, group therapy is certainly more affordable than individual treatment. But does it work? In a study of how group size affects the treatment outcome of arachnophobia, researchers found that groups of three or four people fared better than groups of seven or eight. When a couple of one-inch spiders were crawling on their hands at the end of the three-hour session, those people in the small groups reported substantially less fear than did those in the large groups. Moreover, improvements from small-group sessions were nearly as great as those provided by individual therapy.[29]

Taking efficiency one step further, imagine if arachnophobes could eliminate the therapist altogether and pursue systematic desensitization by using a "how-to" manual. Studies of arachnophobes have found that self-directed programs can work—sort of. Improvements are greatest when the instructions are specific to the particular phobia and when the manual is used in a clinic rather than at home. However, even under optimal conditions, self-help methods have a success rate 20 percent lower than that of therapist-directed treatment.[30]

On the other side of the efficiency coin, psychologists have an understandable interest in what makes for a curable patient. With regard to arachnophobes, three qualities have been examined for their contribution to successful treatment. First, those with a lower background level of general anxiety fare better.[31] In other words, if you're the antsy sort, then therapy has to work against both your specific phobia and your chronic tendencies.

Next, therapists might reasonably predict that patients who use the coping strategy of "monitoring" (the tendency to seek out and attend to a threat) would improve more from exposure than those who rely on "blunting" (the disposition to avoid a feared stimulus and associated information). It stands to reason that monitor types would attend keenly to a spider during exposure, allowing more efficient habituation. However, the research findings are mixed.[32] One study found that those who monitored tended to have a greater reduction in anxiety during a session, but they also relapsed more readily. In another study, the coping strategy had no effect on the outcome of therapy, leaving the scientists to speculate that monitor types might intensify their experience to the point that it interferes with the therapeutic process.

Finally, one might guess that arachnophobes with greater imagery ability would have a more intense experience and hence greater treatment success—at least with Wolpe's traditional form of systematic desensitization.[33] It is true that after therapy, arachnophobes have been found to perceive spiders to be markedly less hairy, ugly, and threatening and to be more prone to describe them as light and feathery if not entirely likable.[34] However, imaginative dullards do as well as Dalí-like patients.

While research hasn't precisely specified the features of an ideal candidate for systematic exposure, psychologists have made significant advances in terms of efficacy, or how to structure therapy for the greatest effect. At the core of such refinements is an understanding of how phobic individuals process the experience of multilegged creatures. For these people, learning not to fear spiders or insects generalizes poorly, so a key to success is diversification.

If a therapist desensitizes an arachnophobe through repeated exposure to a single kind of spider, the patient's fear will diminish rapidly, but it may well return within a month. On the other hand, if the therapist uses a variety of spiders during treatment, then the patient will relinquish her fear more slowly, because the range of stimuli creates ambiguity in her learning. However, such a patient is much more likely to remain free of fear a month later.[35]

Not only does varying the spiders improve the durability of change, but also diversifying the physical conditions of treatment matters.[36] When arachnophobes were tested a week after therapy, those who experienced a spider in exactly the same context as their treatment session showed the least fear. Changing the gender or clothing of the therapist, the size or location of the room, or the color or size of the tools used to contact the spider made the encounter scarier. It's not much help to be brave only in a small white room at the end of a hallway while a guy dressed in a lab coat watches you prod a tarantula with a wooden stick. This is why researchers recommend varying the environment during treatment, which can be accomplished simply by spreading therapy over several days.

All of this might make exposure therapy sound formulaic, but it most assuredly is not. Even the pros make mistakes, such as one case in which a therapist moved a patient too hastily through systematic desensitization. Rather than the woman losing her fear of spiders, the trauma gave rise to a second-order anxiety. At the end of the day, she feared both spiders and the building where the treatment took place. With more therapy, however, she managed to overcome both of her fears (it's not clear whether she was charged for the extra sessions).[37]

Then there was the study in which the researchers planned for the final step of desensitization to involve the patient's holding a spider. The protocol was modified to end instead with the patient touching a spider with a cotton swab, because "during experimenter training, the spiders had occasionally become aggressive."[38] In other words, getting an arachnophobe to finally pick up a spider only to receive a nasty bite would not be a good way to help her overcome her fear. And if the unfortunate person got nipped during group therapy, the other patients might start thinking and understandably head for the door. In fact, the whole matter of thinking about our fears takes us to the next class of therapies.

Let's Think about This . . .

"Now then, Mr. Dalí, you saw Gala hold the grasshopper and she was not harmed, was she?" The artist looked apprehensive. He smiled feebly at his lover and muse as she returned the insect to its jar. The therapist repeated her question, and Dalí replied, "No, I suppose not." The therapist continued, "That's right. So now I'd like you to move halfway to the grasshopper's cage." Gala nodded encouragement as he slid his chair toward the insect on the table.

The underlying assumption of cognitive therapies is that our emotional excesses can be checked by our rational capacities. Faulty thinking got us into our phobias, and clear thinking can get us out (with the help of a therapist).[39] Although how the entomophobe feels (sweaty palms, pounding heart, rising panic) becomes the focus of experience, these emotional and sensory reactions are symptoms. The cause is the consciousness being flooded with mistaken thoughts and doomsday images that build on themselves. Fix the head, and the rest of the person will follow.

A common form of cognitive therapy involves modeling, which is often combined with a version of exposure. Rather than just putting a patient into proximity of a three-inch Madagascar hissing cockroach, a therapist might first allow the entomophobe to watch another person—often the therapist, a parent, a friend, a respected peer, or perhaps a lover-muse—handle the chunky insect without evincing fear or experiencing harm. At least it works for arachnophobes. Simply observing another person handling a spider or even watching a video of someone doing so can yield substantial reductions in fear even after a single session.[40]

Faced with explicit evidence that spiders don't inflict catastrophic harm, the arachnophobe is required to reason through the foundations for her fear. Although such reassessment can be effective, cognition is something of

Figure 7.2
The ultimate accomplishment of the arachnophobe is to physically engage the object of fear—an actual spider. Some therapeutic success accrues from merely watching others handle spiders (modeling) or from seeing images of spiders, but contact with the living creature (in vivo experience) enhances the efficacy of exposure, desensitization, and cognitive therapies (image from Eggybird through Creative Commons).

a double-edged sword when it comes to modeling. The relatively low rates of success with modeling alone appear to result from the arachnophobe's rationalizing that if *she* had handled the spider, the story would not have had such a happy ending.

The Therapist's Swiss Army Knife

Dalí twisted his mustache fretfully as the therapist instructed, "Anxiety is normal. Focus for a moment on Gala." The artist slowed his breathing as the therapist continued: "Take a moment to see how your thoughts about the grasshopper generated your feelings. What has the grasshopper done—and what are you doing to yourself?" The artist admitted that the tightness in his chest was of his own making and edged toward to the insect, murmuring a mantra to build confidence: "Face what you fear; this will pass." When it seemed he could get no closer, the therapist told him to become an actor. Dalí's eyebrow arched inquisitively. "Act as if you're not afraid." Challenged in this way, he came within inches of the insect and felt empowered by his incremental success.

The rock star of therapies for specific phobias is undoubtedly cognitive behavioral therapy (CBT), which is sort of a "best of" album for psychological treatments. This approach combines the evocative emotive capacity of exposure therapy with the compelling rational power of guided, critical analysis.[41]

In the framework of CBT, we might think of a phobic response as a kind of stepping-and-tripping process.[42] For the entomophobe, the first step is an appraisal of the situation. He considers whether the situation—say, a grasshopper on the picnic blanket—represents a clear and present danger. His answer is unconditionally affirmative; the insect is a serious threat to his well-being. In effect, he has stumbled on a misappraisal.

The next step involves drawing upon a "cognitive set"—a way of thinking about the danger that allows him to exclude irrelevant information and rapidly develop a response. Once any of us has engaged a cognitive set, that mode of thinking persists until the danger has passed. However, the entomophobe becomes trapped in a cognitive set because he selects from his memory experiences that affirm the threat (a grasshopper once jumped at him), engages in dichotomous thinking (there is only imminent catastrophe or complete safety), and lacks habituation (when the insect ambles into the grass, there is a sense that the creature is still a threat).

Any number of cognitive sets are available, depending on the degree of threat, but our hypothetical entomophobe might select a "vulnerability set." Within this framework, he underestimates his capacities (he can't conceive of himself brushing away the grasshopper or simply ignoring it), magnifies potential results (if he were to approach the insect, it might dash up his pant leg), and catastrophizes consequences (once inside his clothes, the grasshopper could crawl into his body).

Tripped up by his choice of a cognitive set, the entomophobe next engages in steps that take the form of a dysfunctional dance. A vicious cycle ensues, with his sense of being threatened leading to intense anxiety that feeds back to bolster the perception of danger. The fellow spirals into irrational behavior that an observer might take to be the problem. However, his flailing and screaming are simply the outcomes of a series of cognitive missteps.

* * *

A CBT session is highly pragmatic. The goal is to achieve functionality, not delve into arcane theories regarding the origin of the phobia. As a therapist put it, "One patient had six years of psychoanalytically oriented therapy, which gave her much insight but little help in managing her anxiety."[43]

Rather than recalled childhood events, imagined scenarios, dry labs, or computer simulations, at some point the patient will encounter an actual insect (e.g., Dalí's grasshopper) and experience genuine fear. Along the way, the individual is asked to employ the rational self in challenging the emotional self's theory of catastrophe. In effect, the overarching principle of CBT is to turn the patient into a scientist whose mind and body are the subjects of detached inquiry.[44]

The therapist functions as a trusted professor, structuring a series of empirical experiments (the exposure element) and directing the "student" to draw reasoned conclusions about the results (the rational element). And like a good professor, the therapist uses the Socratic method to lead the patient to reflect on his thoughts and discover for himself the implausibility of his hypothesis of disaster. Data interpretation typically begins with recognition of the automatic thoughts that seem to come from nowhere—the cascade of mental errors that culminate in debilitating fear.

The therapist provides direction and structure, but the patient must ultimately conduct the experiments. Therapy consists of empirically testing reality through engagement with the feared object. The patient is taught to consider his beliefs as hypotheses that are to be revised based on actual results. In effect, he practices the experimental method, asking, "What is the evidence for or against my hypothesis that grasshoppers are dangerous?" The rational self observes the emotional self and ascertains whether fear is justified. As such, it is important that the early investigations generate useful results in terms of testing—and refuting—the phobic hypothesis.

Therapy entails doing, which is the behavioral component of CBT. Even in the initial "experiment," the patient must experience fear and anxiety, not merely imagine it. In this way, it is possible to begin the process of developing an appropriate cognitive set from which viable actions emerge. Along the way,

Figure 7.3
In cognitive behavioral therapy, the therapist combines modern psychology with an ancient teaching technique—the Socratic method. In this eighteenth-century painting by Nicolas Guibal, Socrates is questioning Pericles to draw out what he knows. Likewise, a CBT therapist systematically asks the patient to assess critically whether the patient's fearful beliefs are supported by experiential evidence (image by Rama through Creative Commons).

various methods are used to facilitate this change. Some patients may find considerable relief in learning about the biology of anxiety, so that the symptoms are less mysterious and the feeling is understood to be appropriate and functional in certain circumstances. Reductions in anxiety can also be accomplished through tactics such as distraction (focusing on external features of the world can break the spell of a self-created, internal nightmare) and relaxation methods (allowing the individual to be the master, rather than the victim, of his symptoms).

Once the initial experiment provides preliminary evidence that fear is unjustified, the therapist takes the entomophobic patient through incrementally more challenging tests of the phobic hypothesis. The patient is asked to consider alternative ways of perceiving the situation. With anxiety diminishing at each step, cognitive restructuring leads to a new theory of insects. As the experiments become more demanding, the patient is provided with techniques to avoid having the emotional self jump to unfounded conclusions. The rational self is taught to use methods such as self-instruction, in which calming phrases (e.g., "This feeling will pass" or "Face what you fear") dispel the sense of imminent doom. Or the individual may employ techniques of self-empowerment. For example, as the emotional self is preparing to bolt from the room, the rational self takes over and the patient acts as if he is not afraid—and finding that no catastrophe befalls him, he develops confidence that he can confront the grasshopper without recourse to panic. At some point he might be asked to consider the worst possible outcomes. What if the grasshopper really did land in his hair or crawl across his lips?

With practice and continued "research" into what happens in proximity of the stimulus regarding whether an emotional response is supported by the evidence, the entomophobe progresses from an initial relief of symptoms (slowing of a pounding heart) to the logical analysis of unreflective hypotheses that had been generating those symptoms (grasshoppers will attack me) to the rejection of false hypotheses (grasshoppers do not attack) to the development of justified beliefs (grasshoppers are harmless). And as with any educational setting, there is homework to reinforce what has been learned. The key is to transform the entomophobe into a practicing scientist (without interfering with his artistic genius), able to examine the foundations for his beliefs such that the approaches used in the clinic/"classroom" become a way of life—a way of knowing himself and the world.

Entomologist, Heal Thyself

As for my own grasshopper-induced descent into panic, I managed a kind of intuitive self-treatment that resembled CBT. Not long after my horror-filled encounter on the prairie of Wyoming, I took the opportunity to test my mettle while conducting research in Australia. By this time I had reasoned my way into understanding that my fear was irrational. As an entomologist, I knew that the insects were not a threat to my well-being—physical or psychological. So I asked a colleague to take me into a swarm of Australian plague locusts to see if I was cured (he thought I just wanted to take photographs, as I was too embarrassed to reveal my actual reason).

The Australian locusts form "rolling swarms" that look like shimmering dust storms along the burnt red landscape. Being engulfed by millions of insects evoked a powerful series of emotions, and their movement generated a kind of vertigo. I was mesmerized and astounded by the unfathomable surge of life. However, I was not terrified or even much frightened. There was an enchantment with such wild profligacy that accelerated my heartbeat and breathing, but the experience was much more akin to wonder, a reverence tinged with darkness rather than a nightmarish malevolence.

My approach to grasshoppers based on a form of self-treatment that would be familiar to therapists turns out to have been similar to the psychological tactics taken by others who make their livings immersed in insects.

NOTES

1. Stewart Agras, David Sylvester, and Donald Oliveau, "The epidemiology of common fears and phobias," *Comprehensive Psychiatry* 10 (1969): 151–56; Charles G. Costello, "Fears and phobias in women: A community study," *Journal of Abnormal Psychology* 91 (1982): 280–86.
2. David H. Barlow, *Anxiety and Its Disorders: The Nature and Treatment of Anxiety and Panic*, 2nd ed. (New York: Guilford Press, 2002), 303–7.
3. Ibid.
4. Ibid., 310.
5. Ibid., 380; Lars-Göran Öst, "One-session treatment for specific phobias," *Behaviour Research and Therapy* 27 (1989): 1–7; Joshua D. Lipsitz, Salvatore Mannuzza, Donald F. Klein, Donald C. Ross, and Abby J. Fyer, "Specific phobia 10–16 years after treatment," *Depression and Anxiety* 10 (1999): 105–11.
6. Barlow, *Anxiety and Its Disorders*, 394.
7. Ibid., 417.
8. Phillip Weinstein, "Insects in psychiatry," *Cultural Entomology Digest*, no. 2 (1994), Insects.org, http://www.insects.org/ced2/insects_psych.html (accessed October 31, 2011).
9. Barlow, *Anxiety and Its Disorders*; Aaron T. Beck, Gary Emery, and Ruth L. Greenberg, *Anxiety Disorders and Phobias: A Cognitive Perspective* (New York: Basic Books, 2005).
10. Steve Tromans, "Case notes from therapy," Just Be Well Hypnotherapy, http://www.justbewell.com/phobia-fear-cockroaches.html (accessed March 24, 2012).
11. Barlow, *Anxiety and Its Disorders*; Beck et al., *Anxiety Disorders and Phobias*.
12. Beck et al., *Anxiety Disorders and Phobias*, 127–28.
13. May R. Berenbaum, *Bugs in the System: Insects and Their Impact on Human Affairs* (New York: Helix Books, 1995); Tad N. Hardy, "Entomophobia: The case for Miss Muffet," *Bulletin of the Entomological Society of America* 34 (1988): 64–69.
14. Peter Muris and Harald Merckelbach, "Treating spider phobia with eye-movement desensitization and reprocessing: Two case reports," *Journal of Anxiety Disorders* 9 (1995): 439–49; Peter Muris, Harald Merckelbach, Hans van Haaften, and Birgit-Mayer, "Eye movement desensitization and reprocessing versus exposure *in vivo*: A single-session crossover study of spider-phobic children," *British Journal of Psychiatry* 171 (1997): 82–86.
15. Berenbaum, *Bugs in the System*; Hardy, "Entomophobia"; Beck et al., *Anxiety Disorders and Phobias*.
16. Michael Isikoff, "The lawyer and the caterpillar," *Newsweek*, April 17, 2009, http://www.thedailybeast.com/newsweek/2009/04/17/the-lawyer-and-the-caterpillar.html (accessed April 3, 2013); Ewen MacAskill, "Bush officials defend physical

abuse described in memos released by Obama," *Guardian*, April 17, 2009, http://www.guardian.co.uk/world/2009/apr/17/bush-torture-memos-obama-mukasey (accessed May 30, 2009). Rather less reliable sources indicate that American interrogators may have used insects to extract information in other contexts as well. In the course of convoluted hearsay testimony given at a military tribunal in 2007, the father of a Guantánamo detainee alleged that Pakistani guards had confessed that American interrogators used ants to frighten young children. The boys—ages seven and nine—were suspecting of knowing the location of their father: Khalid Sheikh Mohammed, the purported mastermind of the September 11, 2001, terrorist attacks on the United States. According to the detainee's father, "The boys were kept in a separate area upstairs and were denied food and water. . . . They were also mentally tortured by having ants or other creatures put on their legs to scare them and get them to say where their father was hiding." John Byrne, "Bush memos parallel claim 9/11 mastermind's children were tortured with insects," Raw Story, April 17, 2009, http://www.informationclearinghouse.info/article22440.htm (accessed April 3, 2013).

17. Jeffrey A. Lockwood, *Six-Legged Soldiers: Using Insects as Weapons of War* (New York: Oxford University Press, 2009), 328.
18. Barlow, *Anxiety and Its Disorders*, 408–12.
19. Beck et al., *Anxiety Disorders and Phobias*, 210–30.
20. Iris Nelissen, Peter Muris, and Harald Merckelbach, "Computerized exposure and *in vivo* exposure treatments for spider fear in children: Two case reports," *Journal of Behavior Therapy and Experimental Psychiatry* 26 (1995): 153–56.
21. Kathryn L. Smith, Kenneth C. Kirkby, Iain M. Montgomery, and Brett A. Daniels, "Computer-delivered modeling of exposure for spider phobia: Relevant versus irrelevant exposure," *Journal of Anxiety Disorders* 11 (1997): 489–97.
22. Barlow, *Anxiety and Its Disorders*, 408–17.
23. Ibid.
24. *My Extreme Animal Phobia*, promotional material, Animal Planet, http://animal.discovery.com/tv/my-extreme-animal-phobia (accessed March 24, 2012).
25. Öst, "One-session treatment for specific phobias"; Barlow, *Anxiety and Its Disorders*, 408–17.
26. Barlow, *Anxiety and Its Disorders*, 417.
27. Craig N. Sawchuck, "The acquisition and maintenance of disgust: Developmental and learning perspectives," in *Disgust and Its Disorders*, ed. Bunmi O. Olatunji and Dean McKay (Washington, DC: American Psychological Association, 2009); Suzanne A. Meunier and David F. Tolin, "The treatment of disgust," in Olatunji and McKay, *Disgust and Its Disorders*.
28. Peter J. de Jong, Ingrid Vorage, and Marcel A. van den Hout, "Counterconditioning in the treatment of spider phobias: Effects on disgust, fear and valence," *Behaviour Research and Therapy* 38 (2000): 1055–69; Harald Merckelbach, Peter J. de Jong, Arnoud Arntz, and Erik Schouten, "The role of evaluative learning and disgust sensitivity in the etiology and treatment of spider phobia," *Advances in Behavior Research and Therapy* 15 (1993): 243–55; Meunier and Tolin, "The treatment of disgust."
29. Lars-Göran Öst, "One-session group treatment of spider phobia," *Behaviour Research and Therapy* 34 (1996): 707–15.
30. Barlow, *Anxiety and Its Disorders*, 410; Kerstin Hellström and Lars-Göran Öst, "One-session therapist-directed exposure vs. two forms of manual-directed self-exposure in the treatment of spider phobia," *Behaviour Research and Therapy* 33 (1995): 959–65; Lars-Göran Öst, Paul M. Salkovskis, and Kerstin Hellström,

"One-session therapist-directed exposure vs. self-exposure in the treatment of spider phobia," *Behavior Therapy* 22 (1991): 407–22.
31. Peter Muris, Birgit Mayer, and Harald Merckelbach, "Trait anxiety as a predictor of behaviour therapy outcome in spider phobia," *Behavioural and Cognitive Psychotherapy* 26 (1998): 87–91.
32. Barlow, *Anxiety and Its Disorders*, 414; Peter Muris, Harald Merckelbach, and Peter J. de Jong, "Exposure therapy outcome in spider phobics: Effects of monitoring and blunting coping styles," *Behaviour Research and Therapy* 33 (1995): 461–64; Gail Steketee, Shirley Bransfield, Suzanne M. Miller, and Edna B. Foa, "The effect of information and coping style on the reduction of phobic anxiety during exposure," *Journal of Anxiety Disorders* 3 (1989): 69–85.
33. Barlow, *Anxiety and Its Disorders*, 414.
34. S. J. Rachman and M. Whittal, "The effect of an aversive event on the return of fear," *Behaviour Research and Therapy* 27 (1989): 513–20.
35. Melissa K. Rowe and Michelle G. Craske, "Effects of varied-stimulus exposure training on fear reduction and return of fear," *Behaviour Research and Therapy* 36 (1998): 719–34.
36. Susan Mineka, Jayson L. Mystkowski, Deanna Hladek, and Beverly I. Rodriguez, "The effects of changing contexts on return of fear following exposure therapy for spider fear," *Journal of Consulting and Clinical Psychology* 67 (1999): 599–604.
37. A. Keltner and W. L. Marshall, "Single-trial exacerbation of an anxiety habit with second-order conditioning and subsequent desensitization," *Journal of Behavior Therapy and Experimental Psychiatry* 6 (1975): 323–24.
38. Rowe and Craske, "Effects of varied-stimulus exposure training," 724.
39. Beck et al., *Anxiety Disorders and Phobias*, 190–209.
40. Barlow, *Anxiety and Its Disorders*, 410; Lars-Göran Öst, Ingela Ferebee, and Tomas Furmark, "One-session group therapy of spider phobia: Direct versus indirect treatments," *Behaviour Research and Therapy* 35(1997): 721–32.
41. Aaron T. Beck, *Cognitive Therapy and the Emotional Disorders* (New York: Penguin, 1991).
42. Beck et al., *Anxiety Disorders and Phobias*, 37–77.
43. Ibid., 182.
44. Ibid., 169–89 (see also 37–43, 54–86, 96–97, 190–209); Barlow, *Anxiety and Its Disorders*, 415–16.

CHAPTER 8

Overcoming Fear and Disgust for Fun and Profit: The Professionals

People who work in frightening situations must overcome fear to perform their duties. The high-rise window washer cannot freeze, the emergency room nurse cannot faint, and the soldier cannot flee. Studies have shown that such professionals develop "counterphobic" behaviors (fearlessness in the face of things that other people would find terrifying) through two mechanisms.[1]

Some dampen their anxiety by focusing on the task itself, rather than the larger context. The miner attends to placing the blasting cap correctly rather than imagining a cave-in, and the firefighter focuses on searching each corner of the smoke-filled bedroom for a child instead of contemplating an explosive backdraft.

Other individuals overcome their baseline fear by accruing sufficient experience through repeated encounters with scary situations. With practice, the lifeguard comes to understand that the drowning person will not drag him under, and the police officer realizes that she can apprehend the angry drunk.

Investigations of how professionals come to perform effectively in frightening or disgusting situations have focused on people in certain classic occupations, such as emergency personnel, rescue workers, law enforcers, military personnel, and medical responders. But what about those who willingly encounter swarms of bees and hordes of cockroaches? How do apiculturists and exterminators overcome what would be, for most of us, moments of sheer terror? Although no comprehensive investigation has been undertaken, case studies can be very revealing.

DON'T BEE AFRAID

The state patrolman was alarmed by the dozens of stings on the driver's arms and neck. So rather than issuing a speeding ticket, the trooper insisted on escorting the man to the hospital. In a moment of carelessness, the fellow had dropped a frame of bees from one of the less docile hives in the US Department of Agriculture's bee yard—and he had paid the price. Fortunately, there were no medical complications. And surprisingly, there were also no psychological ramifications. The next day, Dick went back to work with the bees.

Dick Nunamaker worked for nearly thirty years at the USDA's Bee Disease Research Laboratory (and its successor, which focused on insect-vectored diseases of livestock), and he taught classes in the biology of the honeybee at the University of Wyoming.[2] For most of his life he has kept his own hives. And although he could readily withstand stinging insects, bumbling bureaucrats were another matter. So Dick retired to Colorado and planted one of the highest-elevation vineyards in the world. "Most years," he says, "I produce an impressive crop of grape leaves," but when his pinot gris produces, it makes an outstanding wine.

Looking at Dick, you likely wouldn't conclude that he is endowed with superhuman courage. His is more the aura (and appearance) of Captain Kangaroo than of Captain America. But put him in the midst of thousands of bees and Dick doesn't flinch. Instead he finds tranquility.

Dick says that he cannot recall ever having a fear of bees, wasps, or any stinging insects. What he does remember, however, is a childhood spent exploring the oak-hickory forest behind his home in Ohio:

> As a youngster, I watched and wondered if the actions of insects were really as purposeful as they appeared to be. Whenever I went into the woods I was more interested in finding a wasp nest than in spotting a whitetail deer. It was interesting watching the apparent chaos in an ant nest that I disrupted with a twig. There would be ants scurrying everywhere for a short time, and then they would soon all settle down as if they each knew what they needed to do in order to restore order.

What he calls the "mystery of social organization" became a siren song—a source of enchantment and attraction. As for the capacity of ants, bees, and wasps to inflict pain, Dick recounts: "Although I knew they stung, the stinging part was never much on my mind." While other kids had baseball pennants and black-light posters, Dick had half a dozen abandoned paper wasp nests hanging in his bedroom.

Dick's father actively encouraged his son's sense of wonder, accompanying Dick on his entomological excursions when work allowed. He provided the usual parental advice, "If you don't bother them, they won't bother you," and the boy took it to heart. His mother was not especially excited by her son's peculiar fascination but she didn't discourage him—and tolerance is tantamount to affirmation when it comes to kids hanging wasp nests in the house.

The term *Peter Pan syndrome* is sometimes used to describe entomologists who developed a childhood fascination with insects and found a way to never grow up. For Dick, studying bees in college fulfilled two desires. He could feed his boyhood curiosity, and entomology provided a distinctive identity. Friends and relatives were intrigued. It also helped that the study of bees—a practical science of agricultural importance—was a respectable field for a young man from a working-class family. It wasn't as if he was studying butterflies. And, in fact, Dick wasn't much interested in six-legged creatures other than the social insects. Throughout his education (he earned a doctorate in entomology with a dissertation on distinguishing subspecies of honeybees) and research career, Dick sustained the capacity for wonder: "Still today, lots of times I see bees actively engaged in 'something' yet I can't determine what they are doing. It's fascinating, yet perplexing. I'd imagine there is as much or more written about honeybees than any other insect, yet there is so much we don't understand."

From a psychological perspective, Dick's childhood was somewhat exceptional. Although lots of kids may have an early interest in insects, entomologically supportive fathers and tolerant mothers are not nearly so common (I was blessed with a similar upbringing—and I felt the same desire to study something unique and still respectable). It was also unusual that Dick's interest was not in throwing rocks at wasp nests but in carefully observing them. This cognitive engagement laid the foundation for his ability to keep his emotions in check when he finally took the leap into beekeeping.

* * *

In effect, Dick used cognitive behavioral therapy to avoid developing a fear rather than to treat an existing phobia. Before he set up his own hive, Dick listened to experienced apiculturists, read dozens of books, imagined every step he needed to take with the bees (a standard technique used in systematic desensitization), and planned for every contingency, including where to place his charges:

> I vividly remember establishing my first hive of bees, with a 3-lb. package, near the corner of Shields and West Vine in Ft. Collins in a small apple orchard owned by a surly old lady we called Granny. She sold me a copper cup for $15 that she said had been in her family for generations. I learned a couple of weeks later, from her daughter, that she paid $5 for it two weeks earlier at a yard sale.

At least he had the bees figured out. For many of us, a fear of insects is rooted in their being so profoundly "other." Dick's intense study allowed him to avoid this impression: "I never perceived them as alien, and apart from having more legs than us and a few extra things such as antennae—which has always made me envious as I would love to have antennae—I don't perceive them as being all that different from us." Indeed, Dick sees them as models of how we might behave in a more civilized and effective manner.

Taking a cue from William Wordsworth's classic poem "The Tables Turned," which exhorts the reader to "let nature be your teacher," Dick contends that bees could mentor humans about how to enact genuine democracy. When searching for a new location to set up house, a swarm gathers information from scouts who pass along their assessments of prospective sites through the famous dance language of bees (various movements are used to communicate the location and desirability of a resource). There is no corner-office executive, handpicked consultant, or self-appointed expert calling the shots. As Dick notes:

> Some corporations are finding that they can make better decisions by involving all the employees in the company as opposed to simply relying upon the information provided by their team of management experts. I find myself continually drawing fairly unconventional comparisons between insect societies and human societies, and I include these in my teaching.

Far from being sinister, otherworldly beings that evoke a sense of dread, the bees are pretty much like us, except more decent.

Dick's path into beekeeping was a variation on a common theme. Many of his fellow apiculturists are third- and fourth-generation beekeepers, having inherited or bought into family businesses. Dick managed to pick the brains of the old hands by visiting with them and reading their advice, exemplifying how knowledge and rationality can set the stage for diminished fear. So when a new beekeeper is stung, the cognitive model works against associating the insect with pain and thereby conditioning a fearful response. At least for Dick, the possibility of being stung does not evoke anxiety; rather, it provides a paradoxical source of calm.

The two sources of greatest tranquility for Dick are fly fishing and working bees. Many of us can relate to the former, but the bees? Not so much. Dick goes so far as to describe beekeeping itself as a kind of therapy, a refuge from problems and worries. Not only does he eschew veils, gloves, and all other protective equipment, but he also wears shorts when he's working (he does use a device that emits smoke that pacifies the bees, who are understandably upset when their hive is opened). For him, it is precisely because bees can sting that he must be fully in the moment: "I'm concerned with making slow, decisive, deliberate movements and not trapping a bee under my fingers or on the inside of my knee as I squat down."

Like fly fishing, working with bees demands complete absorption in the task. If your mind wanders while you're fishing you might miss a strike, and if your attention wanes in a bee yard the insects will provide a sharp reminder to pull you back into the flow. Like a guru carrying a switch while trying to teach

Figure 8.1
Dick Nunamaker, a US Department of Agriculture scientist turned beekeeper and viticulturist in western Colorado, shows a reporter from the local paper that bees can be handled in relative safety when one treats the insects with care and respect. Even when Dick was a child, his fascination with, and admiration of, the social insects trumped any sense of fear (image by William Woody of *The Daily Sentinel*, Grand Junction, Colorado).

an undisciplined student, the bees provide immediate lessons when a pupil's mind wanders. So for Dick, a bee yard is a kind of sanctuary, an unexpectedly peaceful place where anxiety dissolves.

When things are going well with his bees, Dick reports feeling "an overall sense of health and well-being." And he can determine whether the bees are also flourishing by allowing his gaze to pass over the entire colony, not picking out individual bees but attending to the emergent sense of vigor and vitality.

When pressed, Dick admits that he knows the bees "are capable of causing me great pain—or even killing me." But he understands the nature of the risk, rationally grasping the potential costs and the psychological benefits. However, his unconventional approach to self-administered cognitive therapy goes only so far.

* * *

Dick does not admit to full-blown ophidiophobia, describing his fear of snakes as a "general dislike." But his aversion toward snakes is revealing in terms of understanding his counterphobic approach to bees. The same factors that allow him to interact happily with bees are inverted with regard to snakes: childhood experience, subsequent reinforcements, cognitive reflection, and perceived alienation.

Dick traces his dislike of snakes to the childhood jaunts that fed his affinity for bees. While he was hunting for social insects, his dog had another agenda: "Our boxer used to find snakes and then throw them up in the air over and over again, mostly just playing with them. They'd usually end up dying, but she was like some kind of 'snake dog.'" His boyhood aversion grew to include even harmless garter snakes. One might imagine that Dick's training in biology would have undermined his irrational fear, but a traumatic laboratory encounter only made matters worse.

As evidence that nontherapeutic exposure can exacerbate fear, consider Dick's account of an incident that took place during one of his college courses in animal biology:

> In the lab we did an experiment where I presented two water-filled balloons to a rattlesnake. One balloon had warm water and the other had cold water. The idea was to show that the snake would strike the balloon with warm water [because of the snake's heat sensors]. Well, the snake struck the balloon, which immediately broke, pouring water all over the floor. The striking snake startled me and I slipped in the puddle of water. I was sure the snake would jump down and bite me.

This event illustrates the value of combining exposure with cognitive processing. Indeed, the capacity of emotion to rush in before thinking can happen appears to be key to Dick's aversion to snakes—and to spiders. He has always favored rural living, and such settings generate scary surprises:

> I've encountered so many black widows at unexpected times. Like reaching for the handyman jack that's been sitting in my shed for three years and finding a huge black widow on the handle. The element of surprise by a venomous creature bothers me. Perhaps that is what I don't like about snakes. I remember when I was six or seven, I was visiting my cousins who had a dairy farm and a bunch of chickens. I remember snooping around the barn and moving a big barrel only to find a cottonmouth behind it—coiled up and ready to strike! I remember it as though it happened yesterday.

Finally, while Dick perceives bees as familiar, he judges spiders to be "creepier" than insects, and as for snakes: "Hell, I can't discern their body parts! Where does the neck leave off and the tail begin? And there are no arms or feet! What's with that? Seeing a snake and watching its locomotion is creepy to me. And some are venomous which adds to the fear." One might note that bees have six legs, move erratically, and inject venom. However, once emotion has taken root, rationality has a tough time making its case. Ironically, though, Dick seems to have a clear sense of why others dislike his beloved bees and how a bit of desensitization along with cognitive behavioral therapy (although he doesn't use these terms) can relieve their irrational fears.

* * *

Having worked with bees and interacted with the public for decades, Dick has some valuable insights as to the origins of apiphobia (the fear of bees). In his experience, the strands that run through nearly everyone's fear are socialization, pain, and death.

Hollywood set the stage for panic with *The Deadly Bees* (1966) and *The Swarm* (1978). When newspaper headlines declared "Experts Say 'Killer Bees' to Reach U.S. This Spring" in 1990,[3] Dick was busy developing a biochemical method to identify Africanized bees. He remains disgruntled with the media's sensationalism, which was designed, in his estimation, to frighten people. And he knows firsthand how bees are presented to the public.

A few years ago, Dick was called down to Denver to help salvage what was left of eight hundred hives after a bee-carrying semi overturned on the busiest interchange in the city. The overwhelming sense among the passersby, journalists, and even some of the beekeepers working on the cleanup was fear. He recalls the television crews:

> They looked all around as they were reporting, the same way that reporters are nervous and excited when covering a hurricane. I just think so much is about instilling fear as opposed to simply reporting the situation. It makes sense, I suppose, if the objective is to increase the viewing audience and one's ratings.

The reporters hid in their cars, coming out just long enough to talk to the camera and then retreated to the safety of their vehicles. Of course, they were about eight hundred times more likely to die in an automobile accident than from a bee sting. But their message was clear—bees are dangerous.

And bees are dangerous because they can inflict searing pain and even death—or so many people believe. While it is true that it hurts to be stung (that being the whole point of the encounter, from the bee's perspective), Dick describes the experience as uncomfortable, "but not nearly as bad as hitting my thumb with a hammer." And while it is true that bees kill more people in the United States each year than any other animal, the vast majority of these deaths are the result of allergic reactions. Dick notes that while the public estimates that at least 50 percent of people are allergic to bee stings, the actual frequency is less than 1 percent. People may be mistaken, but under the right conditions they are also teachable—as are bees.

It turns out that not only can we be conditioned to fear bees, but bees can also be conditioned to fear us. While working for the USDA, Dick was involved in a project testing various insecticides on colonies of bees. Because the same hives were used every year, the bees associated the frequent arrival of the government vehicle and its human occupants with unpleasantness—sudden disturbances and poisoned nestmates. This was the only time in his life that Dick wore a bee suit, because the insects were so aggressive that they would sting the rubber gasket around the windshield as the Suburban pulled into the bee yard. The situation—apiphobic humans working with anthropophobic bees—was stressful for everyone, and to make matters worse, Dick found the entire research venture to be inane. "After many years of this craziness," he recounts, "the Bee Lab came to the following amazing conclusion: Insecticides kill insects and honeybees are insects. Any questions?" Rather than teaching bees to fear humans, Dick spends his retirement teaching humans not to fear bees.

People often stop to watch Dick out in the field working his hives. As they come close enough to call out to him, their top three questions are "Aren't you afraid?" "How many times have you been stung?" and "Does it still hurt when you get stung or are you immune to the pain?" He explains that bees really do need to be taken seriously because they can inflict damage, but if you understand them and do the right things, then you give the bees no reason to sting (which is lethal to the insect, so it's not a frivolous act on the bee's part). Visitors can see this jovial fellow wearing shorts and a T-shirt, confidently and calmly moving among the insects without any evidence that he's being harmed. The empirical data are contrary to their hypothesis of danger. For that matter, they're only twenty feet away or so, and the bees aren't coming after them either. Firsthand experiences begin to undermine their irrational fear.

What a professor of entomology frames as a "teachable moment," a psychologist might see as a therapeutic opportunity. Given the parallels between cognitive therapy and teaching, it shouldn't be surprising that many of the elements used in the treatment of phobias can be found in Dick's interactions with the public and his students. And modeling is perhaps the most prominent technique.

Dick has found that people are often afraid of bees because they think a sting is extremely painful, so in his classes and workshops he routinely aggravates a bee into stinging him to demonstrate that being stung is hardly agonizing. His other goal in these demonstrations is to show the students that the venom sack continues to pump even when the bee has departed, which is why you can minimize the discomfort when you are stung by immediately removing the stinger by scraping (not plucking, which squeezes the contents of the venom sac into your tissues). But one bee in a classroom is not the same as thousands of bees in the outdoors.

In his most recent class, there were fifteen students who wanted to become hobbyist beekeepers, but only one of them had ever worked with bees. Dick laid down the rules for having a pleasant time with bees: slow and deliberate actions, light-colored clothing, no perfumes or lotions, warm and sunny weather, and a docile colony. With these lessons firmly implanted, they headed to the bee yard. Dick wore his typical cutoffs and short-sleeved shirt while offering everyone veils—only two accepted. The bees were buzzing, and several landed on the students' arms and hands, but nobody panicked and nobody was stung. Dick wasn't surprised; he has never had a student stung during his courses and workshops. Although he is fearless, he is not foolish—as evidenced by a recent interaction he had with a newspaper photographer.

The fellow wanted Dick to put bees all over himself to show that the insects are harmless. Dick made a counteroffer: "You know, although I'm not a professional photographer I am pretty good with a camera. How about we put bees all over your body and I'll take the photos?" The photographer didn't like that idea. Dick's point was that bees are not harmless, and sideshow stunts are not the way to educate the public. The insects do not warrant fear, but they deserve respect. While responsible modeling can function as a kind of cognitive vaccination against incipient phobias, Dick has tapped into opportunities for what a psychologist would recognize as systematic desensitization.

Toward the end of a beekeeping class, a sister of one of his students came to pick up her brother. The group was engaged in the hands-on part of the course, and the arriving woman was clearly both fascinated and frightened. From a distance, she called out, "Can I take pictures?" Dick enthusiastically endorsed this idea and went back to his students, keeping an eye on the new arrival. She snapped a few photos and then Dick encouraged her to come closer so that she could capture the details. The woman slowly and methodically moved toward the beehives, with Dick occasionally offering an affirming word. In less than fifteen minutes, she had the camera and herself inches away from a honey frame as she shot close-ups of the queen and drones loaded with pollen. She was delighted—and so was Dick.

After countless hours working with bees and people, Dick understands the nature of this relationship. But beekeeping is not the most common profession involving interaction between insects and humans. There are at least ten times as many commercial exterminators (now called pest control operators) as there are commercial beekeepers. Killing insects generates twenty-five times more revenue than does producing honey.[4] And a crawl space seething with cockroaches is arguably more terrifying than a hive of bees.

THE EXTERMINATION OF PESTS—AND HORROR

When Heather Story was an undergraduate student at Louisiana State University, she also worked for her father's pest control business. The pay was good, but some days it wasn't enough. Heather recalls doing a treatment for a flea infestation, figuring that she could get the job done and still make it to class. All went well, until "I looked down and saw hundreds of black spots all over my clothes, especially in my knit shirt. At first I thought it was grass seed because the yard had not been mowed in some time."[5] In fact, she was peppered with fleas. She headed home, stripped in the backyard and hit the shower. Heather didn't make it to class, but she went on to earn a degree in entomology—and to return to the family business: Baton Rouge Pest Control, Inc.

Heather is a strong, confident, blue-eyed, attractive-but-not-prettified, thirty-something woman. With the right clothes, she'd fit the role of a ranch hand in a Wyoming cowboy bar. I don't know if she drinks, but if she does it's not chardonnay (I'd bet on PBR or Jack Daniels). Heather has the kind of swagger that suggests she's not the sort to be messed with by smart-ass men, girly women, or entomological vermin. I admire her (and her dad, who was my dissertation adviser).

Like Dick Nunamaker, Heather developed a positive relationship with insects during childhood. Her father, a professor of entomology at LSU, was always pointing out insects and telling her about their biology. My kids can attest to the annoyance that comes with having an entomologist for a dad (there's also a cool factor on career days that doesn't accrue to accountants and insurance agents). Heather recalls: "When I was a young child I would always try to go with my dad to his lab so I could see the insects and scientific equipment. On family vacations to the mountains we always looked for insects, plants, and animals to talk about." In short, she grew up with insects and never felt repulsed by these creatures. In fact, while some girls harbor nostalgia for Barbie dolls or frilly dresses, Heather fondly remembers the fireflies of her youth. As a child she marveled at dragonflies and still has a soft spot for these insects. However, as Paul wrote in his letter to the Corinthians (13:11): "When I was a child, I spoke and thought and reasoned as a child. But when I grew up, I put away childish things." Heather's path into pest control was not the softhearted journey of an enchanted child:

> Originally what drew me to pest management was the thought of having one job that paid the bills instead of three that still didn't do it very well. I started in pest management when I was an undergraduate. I didn't really know what I wanted to do for a career at the time.

Although her entry into the field was unromantic, with time she found that being an exterminator allowed her to tap into a youth in which insects were objects of fascination rather than fear:

> At first, clients would ask me questions about insects and I would have to ask others or look up the information and get back to them. I decided that I would take an entomology class so I could answer questions better. Then I took another class and another. In the end, I wound up with a master's degree in entomology.

And from there, she took up the family business with a deeper appreciation and understanding of insects. This cognitive context provided Heather with a further bulwark against negative responses to the sorts of encounters that would send most of us to a therapist. In the case of a pest control technician, there's no time for counseling when roaches are pouring out from beneath a kitchen sink or bedbugs are scrambling for cover from behind a headboard.

Heather admits to a sense of revulsion when she is faced with swarming, skittering, seething insects in tight spaces, so it's not as if she no longer experiences what we might feel. Rather, she has developed a toolbox of mental tactics that bear a striking similarity to the methods used in the treatment of entomophobia.

* * *

Heather's first line of defense is desensitization through repeated exposure. After several thousand hornets have dive-bombed, a few million fleas have leapt, and countless cockroaches have skittered, it takes a great deal more stimulation to evoke fear or disgust in her than in your average homeowner. And when a situation crosses a threshold (she has found that those bean-sized German cockroaches have a proclivity for dropping down a shirt collar,

Figure 8.2
Really big cockroaches such as those shown here seem particularly frightening, but it is the little ones that can work their way into an exterminator's hair and clothing. Heather Story deals with these repulsive insects using several methods of her own design that would be familiar to therapists, including desensitization, refocused attention, positive self-instruction, and critical reasoning (image by ric_k through Creative Commons).

scampering up a pant leg, or wriggling into hair), Heather's next tactic resembles an approach advocated by therapists during exposure therapy. She focuses her attention on something positive or distracts herself with pleasant thoughts, such as the paycheck she's earning for her honest day's work.

Heather is extremely intelligent, so it's not surprising that her most potent tool for fending off a rising sense of fear or disgust is rationality: "I think about how silly my feelings are over harmless insects, and I tell myself that the faster I put my irrational feelings aside the sooner I can get on with the job." This sounds like the mantra of a patient undergoing cognitive behavioral therapy (positive self-instruction is a common tactic). Reasoning is a potent defense against unreasonable fear, and as appalling as a houseful of vermin might be, it actually poses little danger—when you think about it.

In the course of a CBT session, a therapist might help an entomophobic patient get past a particularly difficult point by having the individual imagine "What if I had no fear?" and then act as if this were true to cultivate self-confidence. Or as Heather puts it:

> I did experience disgust early in my career but I knew I had to at least look professional so I swallowed my disgust and just got the job done. Even now, some German roach jobs are really bad and I can't help but feel disgusted. But being a pest control operator, you just have to get the job done and try to not show revulsion.

Heather also contextualizes the nature of her work, much as some first responders set aside their own feelings to focus on those in need. She contends that "people in this field like working with others and helping them live better lives." Few of us think of pest control as being one of the "helping professions," which traditionally include medicine, psychology, counseling, social work, education, and ministry. Teaching a child to read, suturing a wound, and advising unemployed workers are compassionate endeavors. But I can attest to the improved quality of life that comes from exterminating the fleas in your carpet or the roaches in your kitchen (my wife and I lived in an aging mobile home a few blocks from the LSU campus while we were in graduate school, so I know about six-legged infestations).

However, even with years of experience, positive framing, and cognitive sophistication, an exterminator is still capable of feeling horror. Heather admits that although she can admire spiders from a distance, she has no affection for these creatures. She recalls once inadvertently walking face-first into an enormous banana spider web, much to the amusement of the client. By her own admission, Heather's response was rather animated: "I whipped off my backpack sprayer and ran around the backyard trying to swat the spider I just knew was sitting on my head!" In her defense, a big banana spider has a leg spread approaching the size of a human hand, and there's a photographic

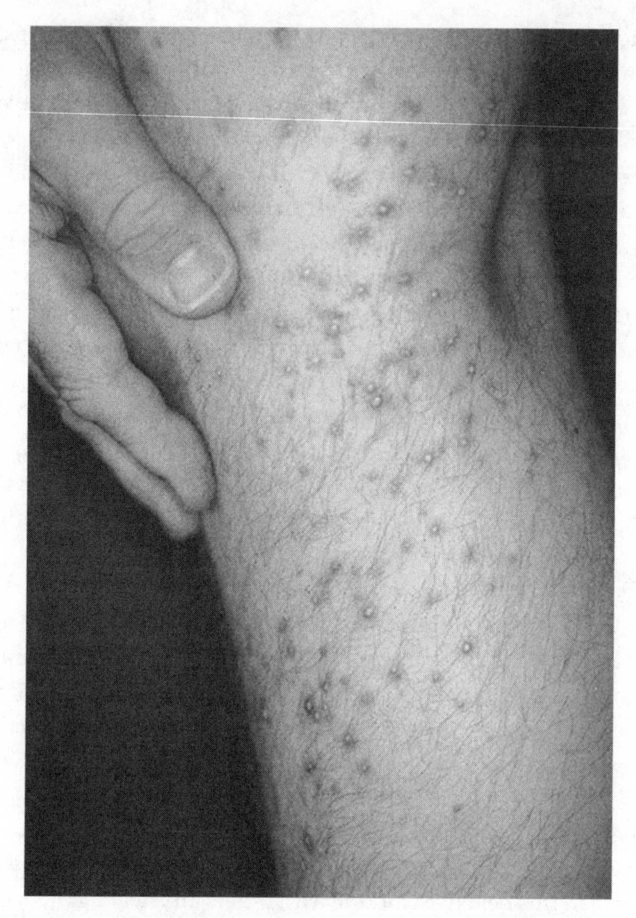

Figure 8.3
Pest control can be understood as one of the helping professions. Just as a firefighter sets aside fear to render aid, an exterminator might deal with a horrific insect infestation by focusing on the quality of life being provided to the client. As for the latter, this image shows the consequences of more than 250 fire ant stings that occurred in less than ten seconds (image by Daniel Wojcik, US Department of Agriculture).

record of one of these creatures dining on a finch that became entangled in its web.[6] If such encounters had Heather on edge about spiders, the deal was sealed with an event that took place in her early twenties.

Wolf spiders have a very disturbing quality (in addition to their typical spidery features)—they can "explode" into hundreds of spiderlings. The impression is generated when a mother spider is swatted while she's carrying a batch of young on her back or a load of little ones still wrapped in an egg sac that she hauls around. The effect, which has been captured on video, is stunning even to a seasoned entomological veteran.[7] Heather learned this the hard way when she was peacefully sitting on the couch one night and suddenly realized that a large brown spider was sitting on her shoulder. When she did the natural thing and smacked the intruder, it instantly exploded into a riot of baby spiders madly scrambling over her and the furniture.

Although the experience deepened her aversion to spiders, Heather's fear does not entail malice: "I felt guilty after I killed the mother, but it was a knee-jerk reaction." Her empathy even for uncharismatic creatures also makes rodent control jobs distasteful. Setting aside the filth and stench, "I am a softie . . . I know they have to be killed but I also think they are cute. I don't really like killing mammals but it's part of the job." She also rescues lizards from sticky traps that her clients use to exterminate these harmless creatures.

When pressed to describe what it is about spiders or insects that elicits fear, Heather doesn't talk about their alien forms and lives, mindless autonomy, or evasive tendencies. Even hordes of insects don't trigger her aversion—unless this is combined with the key property: bodily invasion. "German roaches en masse can make my skin crawl, but only when they crawl on me. I get squeamish when I feel their legs racing over my skin." She's particularly perturbed when she is called in for bed bug treatments—or to control any sort of blood-sucking insects. It's the physical intimacy that is so disturbing, she says, although she pushes her way through the fear and disgust: "The idea of having my personal space invaded by insects en masse is revolting, especially the idea in the back of my mind that I may not even see many of them."

* * *

While Heather's fear and disgust have been distilled to a single feature of insects, her clients' aversions are more wide-ranging. For many of them, lizards evoke terror, as they are perceived as being some sort of snakelike things. And the misunderstanding of insects is even more incredible to Heather: "Believe it or not many people are convinced that lady beetles will bite and sting."

Heather has found that "most people don't like cockroaches because they are 'greasy' or gross looking. They don't like the spiny legs either. I hear a lot of people say they don't like all the legs and hairs." And a horror-fueled imagination can be a powerful engine of irrationality:

> I have clients who get very upset if they see a roach *outside*! They will call me back for a re-treatment because they saw a roach on the porch. People think if they see one roach there must be hundreds or thousands, and the mind just keeps going with some people. I have many people call me back for re-treatment because they are seeing dead roaches!

And then there are the stinging insects, where Heather's world overlaps with Dick's: "Many people fear bees or wasps or any insect that looks like a bee or wasp. I have had clients ask me to spray all the flowers on the property to kill all the bees and wasps that visit them."

All of these emotionally charged responses to insects provide Heather with one of the most rewarding aspects of her job—the opportunity to educate people. In the world of entomophobia, we might think of the pest control operator as the first responder, a kind of intuitive therapist who wants others to lead full and pleasant lives, which sometimes means spraying for vermin but often means helping people live with insects. Heather's own experience with self-administered cognitive therapy puts her in a good position to help others work through their irrational fears:

> Most people can overcome aversions if I educate them about the insects and assure them that the beast in question does not bite or destroy the house. Telling people something interesting about the insect—what it feeds on, how it survives, some evolutionary story—usually works. But there are times when emotion trumps intellect and I regret having to kill insects that should be left alone.

Baton Rouge Pest Control, Inc., seems to be flourishing, even with its technicians talking people out of expensive treatments. A backyard with some buzzing bees, a few garden spiders, the occasional cicada killer (a harmless, two-inch-long wasp that freaks out people more than any other single insect, in Heather's experience), and no fire ants seems like a rational goal. Heather finds genuine fulfillment in "helping people create the environment they like to live in." And for her, that's an environment with humans who understand that most insects are harmless—and do not deserve our psychic venom or chemical poisons.

Heather Story's keen interest in not merely evoking a benign tolerance of insects in her clients but also eliciting genuine fascination with these creatures suggests a radical departure from entomophobia. Heather is an unexpected member of a devoted cadre of entomologists and psychologists who believe that humans are (or can become) entomophiles—lovers of insects.

NOTES

1. Aaron T. Beck, Gary Emery, and Ruth L. Greenberg, *Anxiety Disorders and Phobias: A Cognitive Perspective* (New York: Basic Books, 2005), 127; Seymour Epstein, "The nature of anxiety with emphasis upon its relationship to expectancy," in *Anxiety: Current Trends in Theory and Research*, vol. 2, ed. Charles D. Spielberger (New York: Academic Press, 1972); Isaac M. Marks, *Fears and Phobias* (New York: Academic Press, 1969); S. J. Rachman, *Fear and Courage* (San Francisco: W. H. Freeman, 1978).
2. Author e-mail interview with Richard Nunamaker, August 16 and 19, 2011.
3. "Experts say 'killer bees' to reach U.S. this spring," *Eugene Register-Guard*, April 15, 1990, http://news.google.com/newspapers?nid=1310&dat=19900415&id=g1ZWAAAAIBAJ&sjid=z-sDAAAAIBAJ&pg=5511,3511081.

4. "Honey industry facts," National Honey Board, http://www.honey.com/newsroom/press-kits/honey-industry-facts (accessed April 4, 2013); Heather Collura, "U.S. losing bees and beekeepers," *USA Today*, April 9, 2008, http://usatoday30.usatoday.com/news/nation/2008-04-08-beekeepers_N.htm (accessed May 21, 2012); "Member Directory," National Pest Management Association, http://www.npmapestworld.org/directory/index.cfm (accessed May 21, 2012); "Industry Fact Sheet," National Pest Management Association, http://www.npmapestworld.org/news/factsheet.cfm (accessed May 21, 2012).
5. Author e-mail interview with Heather Story, December 5, 2011, and April 11, 2012.
6. Bonnie Malkin, "Giant spider eating a bird caught on camera," *Telegraph*, October 22, 2008, http://www.telegraph.co.uk/earth/earthnews/3353693/Giant-spider-eating-a-bird-caught-on-camera.html (accessed May 21, 2012).
7. "Pregnant spider explodes with babies when it gets squashed," video, 22 Words, January 18, 2012, http://twentytwowords.com/2012/01/18/pregnant-spider-explodes-with-babies-when-it-gets-squashed; "How to get rid of wolf spiders," CommonPests.com, http://www.commonpests.com/how-to-get-rid-of-wolf-spiders (accessed May 21, 2012).

CHAPTER 9

The Infatuated Mind: Entomophilia as the Human Condition

> Though we welcomed all spiders [in our house] there was the purely practical issue of how difficult it might be to avoid accidental contact with Black Widows. . . . We decided to relocate the youngsters to safer quarters—for them as well as us. We then talked to the mother about her children and told her why we were evicting her offspring.[1]

This passage might seem daft, even deranged, to an arachnophobe, but it is heartfelt. I can vouch for its authenticity because I know its author well. Eve Schnoeker-Shorb and her husband Terry, who shared the spidery home in Arizona, have been friends of mine for years. Eve is a spindly (even spidery) woman of great sensitivity and intensity. And her account of living with spiders is vivid and authentic. Eve's description of her relationship with Goldman (a cellar spider named for her color) is particularly illustrative:

> I soon learned to discern Goldman's temperament, and this through her bouncing ability [cellar spiders bounce in their webs to evade predators]. Goldman had her own bouncing code for me. A slow bounce seemed to be a friendly communication. . . . A medium-paced bounce meant that she did not want to be disturbed. . . . And a frantic bounce indicated that my presence really irritated her—and this was particularly so during the time span that her abdomen was developing in the shiny, bloated state right before she produced her egg sac.[2]

And when the eggs hatched:

> Those babies that did not disperse to other rooms of the house had found appropriate the formation of a triangular sheet web that connected the tips of our toothbrushes to the bathroom mirror and to the mother web.[3]

The crowding of humans and spiders in the bathroom led to some compromises so that Eve and Terry could use the sink and the spiders could have their silken space. And the Shorbs' rearrangement of their lives and their acceptance of multilegged creatures was not limited to spiders:

> Because we were both inclined toward fascination with creatures, especially those that crept and crawled, we were not distressed by the fact that the first residents we encountered at Petunia Manor [their ironically named ramshackle house] existed in great numbers, mostly on the old broken down porch. . . . The earwigs went about their lives mostly outside the manor, but occasionally they ventured onto the floor mattress upon which we slept.[4]

When the time came for Eve and Terry to move, the couple knew that their landlord would surely exterminate his nonhuman tenants. Something had to be done with their little friends, so they designed a "cardboard moving van" that could hold as many as fifty spiders, and after four dozen trips the house was fully evacuated into the surrounding habitat.[5]

What compelled Eve and Terry to act with such anomalous compassion? They viewed the act of relocating the spiders as reciprocity for the spiders' having lifted their spirits. Not only had the spiders provided a source of fascination and serenity for a couple of struggling graduate students, but they had also served as exemplars of "humility, proportionality, living lightly, soft power, and perseverance."[6]

While the creatures provided intellectual, emotional, and spiritual support for the Shorbs, there is a central concept that seems to encapsulate their response: love—or what has been called biophilia. The insects and spiders had transformed their hovel "from a lifeless, bug-bombed out shed to a haven."[7]

Rather than a psychological aberrance, had Eve and Terry tapped into a universal, latent potential? Are we all capable of entomophilia?

* * *

For the past thirty years, the concept of biophilia has generated lively debate among environmental psychologists, conservation biologists, and social scientists. The notion began with E. O. Wilson's simple definition: "The innate tendency to focus on life and lifelike processes."[8] Harvard's don of sociobiology joined forces with Yale's guru of social ecology, Stephen Kellert, to make the case that biophilia increased our ancestors' genetic fitness.[9] But others saw biophilia as more than a product of evolution.

Biology matters, but we are social beings. Eleonora Gullone, founder of Psychologists for the Promotion of the Human Animal Bond, has argued for biophilia's arising from a "biocultural evolutionary process."[10] Perhaps David Orr, a distinguished professor of environmental studies at Oberlin College, puts it best in maintaining that whatever is in our genes, "the affinity for life is now a conscious choice we must make."[11]

A TAXONOMY OF NATURE LOVERS

From the conceptual muddle in which biological evolution and human culture have evidently conspired to make us more fond of forests and kittens (and perhaps grasshoppers) than of shopping malls and pocketknives, we might expect only vague applications to entomophilia (which we'll take to encompass arachnophilia). However, Kellert has provided a compelling typology of nine biophilic values that raises the intriguing possibility of the infatuated—rather than infested—mind.[12]

Utilitarian Affection: What Can Insects Do for Us?

When biophilia manifests in utilitarian terms, we value the practical, material uses of nature. Evolutionarily, this attachment functions to assure physical sustenance and security; we flourished when we attended to living things that met our biological needs. In my work as an entomologist, utilitarian entomophilia emerged through the discovery that some beneficial grasshoppers selectively consume noxious weeds on the prairies and that various predators and parasites provide valuable services in suppressing pestiferous grasshopper species.[13]

Jared Diamond, famed ecologist and environmental historian, takes a hard-line approach to utilitarian biophilia based on his work with New Guineans, who are arguably the best surviving models of prehistoric humanity. Without any sense of curiosity, these people name species that serve a purpose. In Diamond's estimation, humans exploit what their technology allows: "The more susceptible species become depleted or exterminated, leaving less susceptible species which people continue to hunt without being able to exterminate them."[14] Modern American biophilia takes the form of "I love a good steak!"

Across human history and geography, insects have served our interests in utilitarian ways. These creatures have been a godsend through pollination, seed dispersal, decomposition, and biological control of pests. As such, one might wonder why entomophobia is so prevalent and why so many of the most important predators (e.g., frogs, snakes, bats, and insects) elicit aversion. Perhaps the benefits of these creatures are unnoticeable and delayed, while the harms are apparent and immediate. There is, however, at least one obvious and direct benefit of insects.

* * *

Put simply, insects are good eatin'. Although most Americans react to entomophagy with revulsion and the Food and Drug Administration scrupulously limits insects in our groceries, there is no biologically sound reason to exclude them from our diet. The renowned anthropologist Brian Morris contends that "there is no evidence at all that humans have an instinctive dislike of insects as food. On the contrary, throughout human history insects of many kinds have been eaten by human communities, not simply as 'famine' food but as an intrinsic part of their diet."[15] Other scientists go so far as to contend that "humans initially had an insectivorous diet," basing their assertion on the behavior of primates and the findings of Paleolithic archaeology.[16]

The people of every classical civilization seem to have had insects on the menu, including the Chinese, Egyptians, Greeks, Mayans, Persians, and Romans, to name a few. From Pliny and Aristotle to Immanuel Kant and Carolus Linnaeus, the big thinkers were fond of eating insects. Our aversion to entomophagy arguably arose in modern Europe, where being cultured became equated with rejecting primal behaviors.[17]

Despite the aversion to entomophagy in Western societies, anthropologists have documented 2,086 species of insects that are consumed by 3,071 ethnic groups in 113 countries on every inhabited continent.[18] The Bushmen and Hottentots relish ants, the Pedi of South Africa and Shona of Rhodesia are locust lovers, the Pange of Cameroon eat twenty-one kinds of caterpillars, the Thai savor giant water bugs (*maeng dana*), the Japanese consume ricehoppers cooked in soy sauce and sugar (*tsukudanil*), and the Mexicans use agave caterpillars (*gusanos de maguey*) as appetizers, stink bugs (*jumiles*) as chili seasoning, and eggs of water boatmen (*abuahutl*) ground into flour. And even good old Americans who would cringe at a skewer of barbecued grasshoppers or a plate of roasted termites don't hesitate to consume honey, which is, biologically speaking, bee vomit.

All of this entomophilous entomophagy would warm the cockles—and improve the arteries—of a nutritionist's heart.[19] A grasshopper has about the same amount of protein as lean beef, with just a third of the fat, twice the iron, and four times the calcium. In rural South Africa, people collect up to 40 pounds of mopane worms per hour (emperor moth caterpillars living on mopane trees), with the populace gathering 3.5 million pounds per year. Just twenty of these insects provide three-quarters of a person's daily protein needs, as well as the full allotment of calcium, phosphorus, riboflavin, and iron.

Brian Morris's work in Malawi revealed that entomophagy can be crucial when food stores are depleted before the rains come.[20] Although the Malawian people eat more than twenty different insects, consider just the termites. Compared to 3.5 ounces of lean beef, with 24 percent protein, 18 percent fat, and 270 calories, a daily 2-ounce ration of termites provides 35 percent protein, 40 percent fat, and up to 600 calories.

Figure 9.1
Insects have long been an important part of the human diet, providing both vital nutrients (utilitarian value) and culinary pleasure (aesthetic value). Boiled, sautéed, roasted, and dried insects are still commonly found in markets such as these shown in Thailand and China, where cicada nymphs, butterfly pupae, and giant water bugs provide healthful and tasty treats (images by Chrissy Olson and Robert Ennals through Creative Commons).

In addition to nutrients, insects are an energy boon. For example, American Indians gathering locusts from the shores of the Great Salt Lake acquired ten times more calories per hour than they did by hunting big game and nearly three hundred times more than they did by collecting seeds.[21] Speaking of entomophagy in Utah, using simple methods, an entomologist found that it was possible to collect 18.5 pounds of Mormon crickets an hour, representing the caloric content of forty-three Big Macs.[22] Elsewhere, during a productive month villagers in India collect 400 pounds of moth larvae, native people of Zaire gather 3,000 pounds of termites, and a family in Mexico can amass 6,500 pounds of ants.[23] Eating all of those insects could induce dyspepsia, for which other insects might provide a cure.

* * *

In Western history, two competing theories were used in the development of insect-based medicines. One school advocated *similia similibus curantur*, "Let likes be cured by likes." Beginning in the first century C.E. with Pliny the Elder and continuing into the eighteenth century with James's *Medicinal Dictionary*, these folks recommended hairy insects (flies or bees) to cure baldness, oil of earwigs (mixed with the urine of a big-eared hare) for deafness, termites (notably fecund creatures) to enhance fertility, and *Cossus* caterpillars (which exude a whitish, oily substance) to stimulate lactation.[24]

The other school, established by Galen in the second century, held that health required a balance of factors (hot, cold, wet, and dry). The idea was to treat an ailment with an opposing factor. For example, ants were deemed hot-and-dry, so an alcohol infusion with their crushed bodies was prescribed for cold-and-wet ailments such as excess phlegm, listlessness, and impotence.[25] Having one medicine for both the common cold and erectile dysfunction seems like an incredible marketing opportunity.

While theory can be useful, there's nothing like a dose of empiricism.[26] Some insect-based remedies are demonstrably effective (or at least there's a plausible mode of action). Roasted cicadas are used to treat urinary tract maladies, and these insects are high in sodium, which is an antidiuretic. Bee venom blocks sensory nerves and increases blood flow, allowing effective relief of rheumatoid pain. Parents in Prussia and Jamaica fed their children cockroaches, the greasy coating of the insects serving as an effective lubricant for expelling intestinal worms. Finally, perhaps the strangest combination of insect-based medicine, entomophagy, and entomophobia was reported by the novelist Sax Rohmer, who claimed to have cured himself of a fear of spiders by eating one.[27]

In sum, insects can be useful in myriad ways, and becoming aware of their pragmatic value can foster entomophilia. But falling in love because the other individual is an accomplished chef or a good doctor seems to stretch the notion of philia. Let's consider a more intuitively compelling basis for developing an affinity for insects.

Naturalistic Affection: Enchanted by Insects

The naturalistic form of biophilia arises from direct experience. Beachcombing, birdwatching, and asking why grass is green represent this form of affinity. In terms of human evolution, such engagement was important in developing outdoor skills, physical prowess, and mental acuity. When I began my studies of grasshoppers, I started by watching them. I spent long hours walking through various habitats and halcyon days sitting amid these creatures. This quiet engagement fostered both curiosity that fed my imagination and wonder that enthralled my mind.

Many people might be fascinated by a butterfly emerging from its chrysalis or a bee visiting flowers. And knowing that the bee will return to its hive and perform a "dance" to inform its nestmates of a nectar source is simply amazing. This bit of naturalist knowledge came from the work of Karl von Frisch. However, few people know that this Nobel laureate penned a charming book, *Ten Little Housemates*, to show that "there is something wonderful about even the most detested and despised creatures"—including house flies, fleas, bed bugs, lice, cockroaches, silverfish, spiders, and ticks.[28]

My own fascination with insects began in childhood, and the historian Charlotte Sleigh has argued that "the majority of entomologists over the last century have made a point of tracing their interest in insects to an incident during their early youth."[29] Tapping into this innocent wonder may be key to the biophilia of naturalists. E. O. Wilson has described how he found a "cave ant" out in the open in Trinidad and thereby established that the conventional understanding of these insects was mistaken. It was, in his words, "a small quick victory" that would be appreciated by "perhaps a dozen of my fellow myrmecologists."[30] But the triumph was tantamount to a boyhood thrill.

I know how he felt, as I was absolutely delighted after hours of turning over rocks to find a tiny pink cricket of a species that I had long heard about but never seen living among ants. As I drove back home all those years ago, I basked in the satisfaction that Kellert describes: "Immersing oneself in nature can paradoxically produce feelings of both calm and arousal—a heightened awareness coincident with an increased sense of tranquility."[31] Perhaps it takes specialized training to appreciate cave ants and ant-loving crickets, but it doesn't require entomological education to marvel at the migration of monarch butterflies or the mass emergence of periodical cicadas.

If naturalistic biophilia depends on experience and cognition—we encounter creatures and in so doing are motivated to know about their lives—then entomophilia is in trouble today. A study conducted by Kellert showed that while most people surveyed (77 percent) know that insects lack backbones, the majority of the public also falsely believes that insects maintain a constant body temperature (72 percent; actually they are cold-blooded), that spiders are insects (77 percent; hence the basis for combining ento- and arachnophobia/philia), and cockroaches are beetles (89 percent; cockroaches are related to grasshoppers, crickets, mantids, and their ilk).[32]

Our naturalistic ignorance matters because, if Brian Morris is right, the modern revulsion toward insects is due in part to our inability to distinguish one from another.[33] We crudely lump all multilegged animals into an amorphous group that warrants aversion because a few of these creatures can, in fact, harm us. To foster entomophilia, perhaps we need more than childlike curiosity. We need science.

Scientific Affection: Insects as Intellectually Lovable

Although science is putatively objective, Kellert contends that the systematic study of life can evoke biophilia. For our ancestors, an empirical understanding of nature through careful observation had enormous survival payoffs. Not only did such an approach foster intellectual growth, but grasping the limits of nature constrained our proclivity for self-destructive exploitation. As a grasshopper ecologist, I found delight in transforming the biotic confusion of population outbreaks into elegant equations.[34] I came to understand both Newton's passion for finding order in the world and how science fostered his love of the Creation.

Insects are an endless source of scientific inspiration. If you were to put the name of every known organism on Earth in a really big hat and pull out one slip, you would have a better-than-even chance of nabbing the name of an insect. We have discovered about a million species of insects, and most biologists figure that we're only 5–10 percent done with the task.[35]

With that many creatures, there is a staggering diversity of structures and functions. Consider these items from the *University of Florida Book of Insect Records*:

- A walking stick from the Malay Peninsula reaches 21⅞ inches.
- Males of a parasitic wasp are just 5/1000th inch long.
- The aphid *Rhopalosiphum prunifoliae* takes only 4.7 days to complete a generation.
- Adult female mayflies of *Dolania americana* live less than 5 minutes, in which time they copulate and lay eggs.
- A queen ant of *Lasius niger* lived for 28¾ years in captivity.
- A larva of *Buprestis*, a wood-boring beetle, emerged after 51 years inside a timber.
- A midge in the genus *Forcipomyia* beats its wings 1,046 times per second.
- A swarm of the desert locust, *Schistocerca gregaria*, flew from Africa to the Caribbean in 1988, a distance of 2,800 miles.
- Larvae of a fly (*Polypedilum*) found in Nigeria can survive for 5 minutes at either 392 degrees or –454 degrees Fahrenheit.[36]

My own popular science claim to fame was finding documentation that in 1875 a swarm of the Rocky Mountain locust covered an area larger than the state of California and contained 3.5 trillion insects—a record immortalized in *The Guinness Book of World Records*.[37]

Entomology is a source of scientific wonder, but as much as insects seduce the mind, their beauty is what moves the heart of the entomophile.

Aesthetic Affection: An Entomological Beauty Pageant

Beauty can be a powerful pathway to biophilia. Indeed, it has been argued that "biophilia means the love of life. Classically, the ultimate object of love is the beautiful."[38] The spectacular colors of a butterfly in a meadow of wildflowers and the soothing chirp of crickets on a warm summer night transfix our senses. In evolutionary terms, there was an advantage in being attuned to balance, harmony, order, diversity, and the other properties that we associate with beauty. Moreover, aesthetic pleasure is associated with physical healing and mental restoration.[39] Experiencing beauty is good for humans, and I can attest to the sensory richness of insects:

> Grasshoppers are beautiful animals. On an afternoon's walk, I can find twenty or more species. *Dactylotum bicolor* is a garish pink, blue, and black pencil-stub of a grasshopper. Like Jack Sprat and his wife, the sleek, velvety black males of *Boopedon nubilum* are the antithesis of the obese, mottled-brown females. The size of a mouse, the wingless pink-and-green *Brachystola magna* lumbers across weedy fields munching sunflowers. Only a sharp eye can detect *Hypochlora alba,* a ghostly green grasshopper that vanishes as it tumbles into a patch of the cudweed that constitutes its only food.[40]

E. O. Wilson argues that modern humans are fond of our ancestral landscapes. Around the world, we re-create savannas (city parks and golf courses being fine examples) because these habitats provide a primal sense of well-being. Such settings allow us a clear view of our surroundings so that we can detect predators and prey (or muggers and ice cream stands), while clumps of trees provide shade and water promises food and drink.

When it comes to insects, there is a long history of valuing these creatures for their beauty, from the contemporary villagers of Malawi to the French writers of the eighteenth century, who argued that the study of insects "was better than a visit to the art museum, satisfies curiosity better than travel literature, provides spectacles superior to the opera, and stirs up more worthy passions than the romance novel."[41] This sentiment was carried forth by Walt Whitman in *Leaves of Grass*:

> I believe a leaf of grass is no less than the journey-work of the stars,
> And the pismire is equally perfect, and a grain of sand, and the egg
> of a wren.[42]

A pismire is an ant—the term more crudely formulated as *piss-ant*, an allusion to urination and the smell of formic acid released by these insects. And in their classic work *Extinction*, Paul and Anne Ehrlich echo the earlier, literary sentiments:

> Not only does [insects'] beauty, in our opinion, often outshine the *Mona Lisa*, their variety puts postage stamps to shame, their miniaturization far surpasses the best efforts of engineers, and the drama of their existence can compete with those concocted by the best playwrights.[43]

While insects are often appreciated in natural settings, they are also highly valued as cultural artifacts. Who doesn't enjoy the look and feel of silk? Stag and rhinoceros beetles are the epitome of aesthetic (and economic) value;

prime specimens sell in Tokyo for more than $3,000. There are glossy magazines devoted to the breeding and keeping of these insects. But then, the Japanese have cultured insects for centuries, with "bell crickets" being prized for their songs in the 1700s.[44]

Insects have inspired artists in all places and times. These creatures appear in the visual arts of virtually every culture, including the earliest rock art, which depicts bees and honey gathering.[45] Even the performing arts—music, dance, and theater—find inspiration in insects. Of course there are cultural differences, and one's aesthetic appreciation may not extend to a living, two-inch bejeweled beetle on a golden chain, pinned to a woman's blouse.[46]

Speaking of taste, what about those edible insects? Nutrients and calories are fine, but the culinary arts are a source of sensory pleasure. I've stir-fried and roasted grasshoppers, which have a rather pleasant shrimp-like taste. However, my forays into aesthetic entomophagy pale in comparison to Peter Menzel and Faith D'Aluisio's eight-year world tour of insect cuisines. They report that roasted witchetty grubs from Australia are reminiscent of "nut-flavored scrambled eggs and mild mozzarella, wrapped in a phyllo dough pastry."[47] Sounds good! Thai dishes include fried bamboo worms, which taste like salty, crispy shrimp puffs. An Indonesian recipe for bee larvae involves cooking them with coconut oil, garlic, onion, chili, and papaya (which would pretty much make even tofu delectable). When saucer-sized tarantulas from Venezuela are fire roasted, the white meat in the legs and abdomen has the flavor of smoked crab. Really.

Just so that it is clear that these entomophiles are being honest, they also reveal that insects are not invariably palate pleasing. They compare toasted *tanyo kuro* worms from Peru to the "charred part of a well-done hot dog" and describe a Mexican dish made with stink bugs as tasting like "aspirin saturated in cod liver oil with elements of rubbing alcohol and iodine."[48] But even distasteful insects such as cockroaches (being greasy and loaded with uric acid, along with a dash of ammonia) can be sources of entomophilia if we use our imaginations—rather than our senses.

Symbolic Affection: Inspired by Insects

In the symbolic manifestation of biophilia, we are moved by the capacity of nature to represent our thoughts. Presumably our ancestors who could frame their ideas in the familiar terms of nature flourished. Describing the enemy as being "mad as a hornet" prepared a village for battle more effectively (with the residents becoming "as busy as bees") than a lengthy disquisition on the opponent's psychological state. And perhaps those who worked through their hopes and fears using insect-populated dreams achieved mental health (or at least provide rich fodder for modern psychotherapists).

That said, I have long resented the fable of "The Ant and The Grasshopper" as there is much to be said for living in the moment. The ant saved seeds for the winter while the grasshopper enjoyed the summer. But what if a farmer doused them both with insecticide in the fall? Who'd be the smart one then?

If my defense of the grasshopper seems odd, consider the ways in which the lowly cockroach has been elevated to a symbol of sociopolitical empowerment.[49] These insects have become icons of ecofeminism through their primordial earthiness and defiant persistence. In fact, the cockroach has long been an allegorical hero to marginalized people of African and Hispanic descent. Cockroaches are celebrated in Mexican folklore as symbols of successful opposition to colonialism. The transformation of the cockroach into a champion of the oppressed is described in Amoz Oz's novel *Fima* when a Jewish man encounters the insect:

> Suddenly a cockroach came strolling toward him, looking weary and indifferent. . . . Still on his knees, he slipped off a shoe and brandished it, then repented as he recalled that it was just like this, with a hammer blow to the head, that Stalin's agents murdered the exiled Trotsky. . . . He was filled with awe at the precise, minute artistry of this creature, which no longer seemed abhorrent but wonderfully perfect: a representative of a hated race, persecuted and confined to the drains, excelling in the art of stubborn survival, agile and cunning in the dark; a race that had fallen victim to primeval loathing born of fear, of simple cruelty, of inherited prejudices.[50]

While grasshoppers, cockroaches and other disdained insects occasionally represent virtuous qualities, the winner—at least in Western culture—is the Hymenoptera. The ants, bees, and wasps are orderly, hard-working, and loyal. And of all the symbolic symbioses, perhaps none is more richly developed than that between Christianity and the honeybee.[51]

Bees held places of honor in the sacraments. Baptism with honey-milk signified a foretaste of heavenly sweetness. In recognition that worker bees never mate, only candles made of beeswax, symbolizing the virginal body of Christ, could be used in the celebration of the Mass. Some priests even took the bees to be communicants, placing morsels of the consecrated Eucharist into their hives.

Bees were worthy of such consideration in light of their purity. The hive symbolized a monastery of sexless, six-legged monks, while the "king" of the hive (early Christians didn't figure that a female was in charge) was an insectan pope. Once people figured out the sex of the head bee, Christ was symbolized by the honeybee queen, necessitating a bit of gender bending. More plausibly, she also came to represent the Virgin Mary, given that nobody had seen her mating (copulation takes place on the wing far above the ground).

Along with chastity, bees represented the virtues of industry, order, frugality, courage, and obedience. Monks kept hives as reminders to give themselves fully to the Church. In the Old Testament, bees also represented fertility and abundance in "a land flowing with milk and honey" (Exodus 3:8). And while honey represented God's generosity, the bee's sting represented his severe judgment.

Even Christian numerology was substantiated by bees. Their three body parts represented the Trinity; their eggs hatched in three days, just as Christ was in the tomb for three days before the Resurrection. The reemergence of bees after three months of winter was interpreted as a sign of immortality. And in this context of everlasting life, swarming bees represented souls migrating toward Divine unification.

The interweaving of bees and beliefs gave rise to a wide range of customs reflecting the importance of these insects. The practice of "telling the bees" was prevalent in Central Europe, the British Isles, and North America. This superstition is poignantly captured in John Greenleaf Whittier's poem "Telling the Bees":

> Before them [the hives], under the garden wall,
> Forward and back
> Went drearily singing the chore-girl small,
> Draping each hive with a shred [shroud] of black
>
> Trembling, I listened: the summer sun
> Had the chill of snow;
> For I knew she was telling the bees of one
> Gone on the journey we all must go![52]

Beekeepers informed their apian charges of all important family events.[53] Failure to do so meant that a newborn might die or a marriage could fail. In Old German, the beekeeper was called *Bienenvater*, or "bee father"; no other animal warranted such fondness. Indeed, in German and French the same verb is used when bees or people die, while all other animals merely perish. So special were bees in the lives of people that one could obtain hives through barter or exchange but never with the crudity of money.

It is difficult to know whether the symbolic power of insects arises from or gives rise to the next manifestation of biophilia. Does Arlington National Cemetery evoke a love of country, or does a love of country compel the keeping of a cemetery for the nation's fallen soldiers?

Humanistic Affection: Brother Grasshopper, Sister Cockroach

Beneath instrumental value lies intrinsic worth capable of eliciting unconditional love. This form of biophilia provided our ancestors with the capacity for bonding, sharing, and cooperating, which was advantageous for social groups. By extending humanistic care to other life forms, the conservation

Figure 9.2
Humanistic biophilia is most familiar in terms of our relationship to companion animals. The concept of agape entails unconditional love in which the other being is not expected to return our affection—a notion that would seem particularly viable in terms of insects. However, the pets in this painting seem more interested in the girl's breakfast than in the child (image by Charles Burton Barber 1894 through Creative Commons).

and protection of nature enhanced our own survival. In my collection of essays *Grasshopper Dreaming: Reflections on Loving and Killing*, I describe my struggle with developing methods to control grasshoppers and suggest why many scientists avoid humanistic affection:

> Rather than sustaining the illusion of objectivity, they could open themselves to a respectful, caring, even loving relationship with the creatures they study. But then they would end up like me—attached to the creatures I kill, with all of the unrest that this entails.[54]

Nor am I alone. One of the greatest animal behaviorists of the twentieth century argued that developing affection for other species is, in fact, essential to science. The Nobel laureate Konrad Lorenz wrote:

> It takes a very long period of observing to become really familiar with an animal and to attain a deeper understanding of its behaviour; and without the love for the animal itself, no observer, however patient, could ever look at it long enough to make valuable observations on its behaviour.[55]

Along similar lines, when E. O. Wilson moved away from his Baptist roots, he did not abandon a deep connection to something beyond himself: "We are in the fullest sense a biological species and will find little ultimate meaning apart from the remainder of life."[56] For him, the subjects of his abiding affection are the ants.

Of the four Greek conceptualizations of love, biophilia best fits with *agape*. At least it's clear that Wilson and I aren't experiencing romantic love, or *eros*. And *storge* is pretty much restricted to affection within a family. Nor does the relationship to ants and grasshoppers seem like *philia*, or the love between friends (the root of *biophilia* notwithstanding). David Orr suggests that humanistic affection is akin to what Albert Schweitzer called "reverence for life"—a sense that one is in the presence of something that is simply valuable, worthy in and of itself.[57] Another formulation of agape is love of the unlovable, such as lepers, pond scum, and cockroaches.

As William Miller argues, love can suspend the rules of disgust when we engage sexual partners, dirty-diapered children, and deteriorating parents.[58] This capacity to see beyond the surface into the marvel of another being is captured in William H. Gass's short story "Order of Insects," in which the main character moves from being repulsed by cockroaches to curiosity, sympathy, and finally a kind of passionate obsession.[59] This last step suggests the possibility that entomophilia can cross into irrational ecstasy.

Joanne Lauck is an environmental educator who has gone off the deep end when it comes to the love of insects. This New Age mystic contends that bees are sensitive to "quantum fields of quarks," that humans can telepathically communicate with flies and cockroaches, and that a mosquito's intentions along with our resentment of its feeding determine whether the bite will be itchy.[60] Her view is that the insects are harmless, sensitive, conscious beings, and our "wilderness self" has a great fondness for them. Linda Hogan, another credibility-stretching writer, explains that by listening to "the true soul of the roach . . . we can gain its secret wisdom, embracing the fertile soil of the chthonic earth."[61]

Having drifted into the terra incognita of insects' dark, earthy souls as a basis for affection, let's set sail for a biophilic approach that returns us to rational terra firma.

Moralistic Affection: Ethical Entomophilia

When biophilia is rooted in moral principles, our concern for nature reflects an understanding of what is right. For our ancestors, ethical guidelines were essential to an orderly and fulfilling life; it is hard to flourish when your neighbor is trying to crush your skull. By extension, limiting our impulses likely paid dividends in terms of sustaining vital natural resources and relationships. In my research and teaching, I have often appealed to moral arguments.[62] How we perceive our relationship to insects shapes our behavior in the laboratory and our control methods in the field.

Of the three classic theories of ethics, the most familiar is utilitarianism: Do the greatest good for the greatest number. What matters in this approach are results—and pain is a bad outcome. The philosopher Peter Singer has argued for including all sentient animals in our utilitarian calculation. And the available evidence suggests that insects can feel at least some kinds of pain.[63] This is why students working with live insects in my teaching labs use anesthesia before procedures that might inflict pain—the inconvenience is minimal, the potential ethical benefits are considerable, and the possibility of producing more compassionate, perhaps even entomophilic, students is compelling.

The second ethical theory concerns the reasons for our actions. In short, we should act in accord with our duties (the flip side being the rights of other individuals). The most basic formulation of this approach is the Golden Rule: Do unto others as you would have them do unto you. It's not clear whether we have duties to insects (or whether they have rights), but I wouldn't want to be sprayed with poison without good reason. And if we are obligated to respect other lives, even those that don't reciprocate, such as newborn babies or comatose adults, then perhaps there is a moral bridge to the insects.

The last of the classical frameworks is the cultivation of the virtues—character traits that, if a person possesses them, lead to doing the right thing. There are lots of candidates for the virtues, but surely compassion, gentleness, and wonder are viable possibilities. And all of these might well entail affection for insects. Aristotle, the first champion of virtue ethics in the Western world, thought that the best way to become virtuous was to practice until these qualities became second nature. And insects provide abundant opportunities to refine our character.

Contemporary thinkers have proposed some alternative ethical systems that might also underwrite entomophilia.[64] E. O. Wilson contends that human nature is such that psychological well-being demands a biophilic response to six-legged creatures. Although biologically enlightened self-interest grounds many people's conservation ethic, this kind of ethical egoism might also lead to another, less kind-and-gentle version of biophilia.

Dominionistic Affection: We Love to Control

Contrary to intuition, biophilia does not necessarily involve a positive relationship with nature. Kellert maintains that our desire to control other life forms can form a strong connection. In evolutionary terms, the ability to subdue other beings was vital to defeating enemies and hunting/domesticating animals. In a broader sense, the physical prowess and courage that came with domination fostered the exploration of new lands and the acquisition of resources. For me, as much as I enjoyed grasshoppers in life-affirming contexts, most of my research involved suppressing these creatures—quashing their outbreaks and wresting control of the grasslands from them on behalf of ranchers. And I was damn good at dominion, developing a method that has been used across millions of acres of rangeland in the American West.[65]

There may be no better example of a dominionistic relationship with nature than our efforts at pest control. Insects destroy nearly half of our food before it reaches our plates. And this means war. Entomology is rife with military metaphors, and nobody does it better than the locust control agencies, with their multinational campaigns, logistics manuals, aircraft sorties, strategic maps, and command structures. Having spent a lot of time with these people, I can also see a kind of admiration, bordering on respect if not quite affection, for a worthy foe. As Kellert notes, "Even the tendency to avoid, reject, and, at times, destroy elements of the natural world can be viewed as an extension of an innate need to relate deeply and intimately with the vast spectrum of life about us."[66]

Insects are sometimes cultured, rather than destroyed, in our quest for dominion. Honeybees and silkworms are more like slaves than enemies. But this metaphor isn't quite right, according to an intriguing analysis of early modern manuals for insect keepers.[67] Rather, bees were seen as industrious workers requiring oversight and management, while silkworms were portrayed as fat and lazy, demanding that humans stuff their bellies. But whether loyal laborers produced honey or corpulent clods provided silk, humans ultimately called the shots.

Speaking of shooting, insects have also been conscripted into our wars.[68] In our efforts to dominate other humans, we have made insects into weapons through a kind of double dominion. For example, during World War II the Japanese killed more than four hundred thousand people with cholera-coated flies and plague-infected fleas. And the US Entomological Warfare Division at Fort Detrick weaponized yellow fever mosquitoes during the Cold War. Dominion? Most assuredly. Biophilia? Perhaps.

Today, our power over the insects has taken the form of genetic manipulations. Scientists are busy inserting genes into plants to make them insecticidal and engineering insects to produce lethally flawed offspring. But perhaps we have an affinity for our own creations; the organisms that do our bidding are connected to us. For better or worse, our biologies are interdependent, and that counts for something.

The greatest obstacles to overcome in our search for domination are often psychological, and surely one of the greatest challenges is our fear. Gaining control of our entomophobia might evoke a kind of entomophilia. The elements of such a case unfolded in 2007, when spiders constructed an immense communal web in a Texas state park.[69] This silken structure covered trees along two hundred yards of trail and was inhabited by millions of spiders. People were horrified, but thousands came to see the "creepy" spectacle. It appears that they sought to psychologically vanquish the spiders. Initially horrified, many of the visitors overcame their aversion—and this is not a bad start toward affection.

However, there is one last manifestation of our connection to the living world that reaches beyond mere dominion into a realm from which biophilia becomes a strangely inverted concept.

Negativistic Affection: Dislike Is Not Dismissal

In the final formulation of biophilia, we are repulsed. According to Kellert, disdain and dread are powerful forms of active engagement. Entomophobia counts as entomophilia because the opposite of love is not fear—both of these entail intense relationality. In human evolution, negative responses were critical to survival. Those who attempted to hug tigers, snakes, hornets, and trapdoor spiders didn't leave many offspring. As for me, the grasshoppers in that gully weren't like the millions of negligible grass seeds or sand grains—the insects were vividly, nightmarishly present.

The public cares about insects—or at least most people are not ambivalent. In a survey of more than a thousand households in Arizona, 84 percent of respondents said that they disliked arthropods in their homes, and less than 1 percent liked them, meaning that only one in six didn't care either way.[70] Kellert's work shows that the vast majority of people strongly dislike ants, beetles, bugs, cockroaches, ticks, and especially biting and stinging arthropods such as mosquitoes, scorpions, spiders, and wasps.[71] People are put off by insects spreading disease, eating crops, multiplying mindlessly, and defying our will. In Kellert's study, scientists were found to be the least negative group, and he surmises that education increases appreciation of, and concern for, insects. But this guess might be a bit too quick, given that even acclaimed biophilic scientists harbor reservations about some life forms.

E. O. Wilson has his doubts about life in the oceans: "The living sea is full of miniature horrors designed to reduce visiting biologists to their constituent amino acids in quick time."[72] And he's not the patron saint of arachnophilia either. Wilson recounts a visit to New Guinea during which he impressed the locals by capturing a snake. Charmed by this tall white fellow's courage, the children sought to impress their visitor:

> One brought me an immense orb-weaving spider gripped in his fingers, its hairy legs waving and the evil-looking black fangs working up and down. I felt panicky and sick. It so happens that I suffer from mild arachnophobia. To each his own.[73]

To each his own? This from the spokesman for biophilia. But Kellert has a point. Wilson didn't just smile and nod, as he might have if the kid had presented a handful of leaves or a sparkly rock. The world's leading conservationist was drawn into a negative relationship with the flailing spider. If we take philia to mean affection and we take affection to mean being affected, then perhaps Wilson's response to a giant spider and my reaction to a swarm of grasshoppers are manifestations of entomophilia.

And so, however you parse human emotions, it looks like we're entomophilic. But as we will see, this psychological inevitability may not be a desirable feature of the theory.

NOTES

1. Yvette A. Schnoeker-Shorb and Terril L. Shorb, *The Spiders & Spirits of Petunia Manor* (Prescott, AZ: Native West Press, 1996), 24.
2. Ibid., 11.
3. Ibid., 17.
4. Ibid., 6.
5. Ibid., 59.
6. Ibid., 54.
7. Ibid.
8. Edward O. Wilson, *Biophilia* (Cambridge, MA: Harvard University Press, 1984), 1.
9. Stephen R. Kellert and Edward O. Wilson, eds., *The Biophilia Hypothesis* (Washington, DC: Island Press, 1993).
10. Eleonora Gullone, "The biophilia hypothesis and life in the 21st century: Increasing mental health or increasing pathology?," *Journal of Happiness Studies* 1 (2000): 293–321.
11. David W. Orr, "Love it or lose it: The coming biophilia revolution," in Kellert and Wilson, *The Biophilia Hypothesis*, 416.
12. Stephen R. Kellert, "The biological basis for human values of nature," in Kellert and Wilson, *The Biophilia Hypothesis*.
13. Jeffrey A. Lockwood, William P. Kemp, and Jerome A. Onsager, "Long-term, large-scale effects of insecticidal control on rangeland grasshopper populations," *Journal of Economic Entomology* 81 (1988): 1258–64; Jeffrey A. Lockwood and Charles R. Bomar, "Consumption of prickly pear cactus flowers by *Melanoplus occidentalis*: A coevolutionary association?," *Environmental Entomology* 21 (1992): 1301–7; Jeffrey A. Lockwood, "Management of orthopteran pests: A conservation perspective," *Journal of Insect Conservation* 2 (1998): 253–61.
14. Jared Diamond, "New Guineans and their natural world," in Kellert and Wilson, *The Biophilia Hypothesis*, 268.
15. Brian Morris, *Insects and Human Life* (Oxford: Berg, 2006), 53.
16. Julieta Ramos-Elorduy, "Anthropo-entomophagy: Cultures, evolution and sustainability," *Entomological Research* 39 (2009): 274.
17. Ibid.; Wilson, *Biophilia*.
18. Ramos-Elorduy, "Anthropo-entomophagy"; Alan L. Yen, "Entomophagy and insect conservation: Some thoughts for digestion," *Journal of Insect Conservation* 13 (2009): 667–70; Morris, *Insects and Human Life*.

19. David George Gordon, *The Eat-a-Bug Cookbook* (Berkeley, CA: Ten Speed Press, 1998).
20. Morris, *Insects and Human Life.*
21. David B. Madsen, "A grasshopper in every pot," *Natural History*, July 1989, 22–25.
22. Ibid.
23. Dana Goodyear, "Grub: Eating bugs to save the planet," *New Yorker*, August 15, 2011, 38–46; Ramos-Elorduy, "Anthropo-entomophagy."
24. May R. Berenbaum, *Bugs in the System: Insects and Their Impact on Human Affairs* (New York: Helix Books, 1995).
25. Ibid.
26. Ibid.
27. Gordon, *The Eat-a-Bug Cookbook.*
28. Karl von Frisch, *Ten Little Housemates*, trans. Margaret D. Senft (New York: Pergamon Press, 1960), 141.
29. Charlotte Sleigh, "Inside out: The unsettling nature of insects," in *Insect Poetics*, ed. Eric C. Brown (Minneapolis: University of Minnesota Press, 2006), 287.
30. Wilson, *Biophilia*, 21.
31. Stephen R. Kellert, *Kinship to Mastery: Biophilia in Human Evolution and Development* (Washington, DC: Island Press, 1997), 93.
32. Stephen R. Kellert, "Values and perceptions of invertebrates," *Conservation Biology* 7 (1993): 845–55.
33. Morris, *Insects and Human Life.*
34. Jeffrey A. Lockwood and Dale R. Lockwood, "Rangeland grasshopper population dynamics: Insights from catastrophe theory," *Environmental Entomology* 20 (1991): 970–80; Dale R. Lockwood and Jeffrey A. Lockwood, "Evidence of self-organized criticality in insect populations," *Complexity* 2 (1997): 49–58.
35. Edward O. Wilson, *The Diversity Of Life* (New York: W. W. Norton, 1993).
36. Thomas J. Walker, ed., *University of Florida Book of Insect Records*, http://entnemdept.ufl.edu/walker/ufbir/chapters/index_subject.shtml (accessed July 25, 2012).
37. Jeffrey A. Lockwood, *Locust: The Devastating Rise and Mysterious Disappearance of the Insect That Shaped the American Frontier* (New York: Basic Books, 2004).
38. Michael E. Soulé, "Biophilia: Unanswered questions," in Kellert and Wilson, *The Biophilia Hypothesis*, 452.
39. Roger S. Ulrich, "View through a window may influence recovery from surgery," *Science*, 224 (1984): 420–21.
40. Jeffrey A. Lockwood, *Grasshopper Dreaming: Reflections on Killing and Loving* (Boston: Skinner House, 2002), 29.
41. Marc Olivier, "Through a flea-glass darkly: Enlightened entomologists and the redemption of aesthetics in eighteenth-century France," in Brown, *Insect Poetics*, 243.
42. Walt Whitman, *Leaves of Grass* (1855; repr., Hollywood, FL: Simon & Brown, 2011), 32.
43. Paul Ehrlich and Anne Ehrlich, *Extinction: The Causes and Consequences of the Disappearance of Species* (New York: Ballantine, 1981), 46.
44. Hugh Raffles, *Insectopedia* (New York: Pantheon, 2010).
45. Morris, *Insects and Human Life.*
46. Sumitra, "Maquech beetles—Mexico's controversial living breathing jewelry," Oddity Central, April 3, 2012, http://www.odditycentral.com/?s=maquech (accessed July 25, 2012).

47. Peter Menzel and Faith D'Aluisio, *Man Eating Bugs: The Art and Science of Eating Insects* (Berkeley, CA: Ten Speed Press, 1991), 18.
48. Ibid., 153, 110.
49. Marion Copeland, *Cockroach* (London: Reaktion Books, 2003); Cristopher Hollingsworth, "The force of the entomological other: Insects as instruments of intolerant thought and oppressive action," in Brown, *Insect Poetics*.
50. Amos Oz, *Fima* (New York: Mariner Books, 1993), 78–79.
51. Elizabeth Atwood Lawrence, "The sacred bee, the filthy pig, and the bat out of hell: Animal symbolism as cognitive biophilia," in Kellert and Wilson, *The Biophilia Hypothesis*.
52. John Greenleaf Whittier, "Telling the Bees," 1858, reprinted at Poetry Foundation, http://www.poetryfoundation.org/poem/174759 (accessed July 25, 2012).
53. Lawrence, "The sacred bee, the filthy pig."
54. Lockwood, *Grasshopper Dreaming*, 43.
55. Quoted in Philip N. Lehner, *Handbook of Ethological Methods* (New York: Cambridge University Press, 1996), 11.
56. Wilson, *Biophilia*, 81.
57. Orr, "Love it or lose it."
58. William Ian Miller, *The Anatomy of Disgust* (Cambridge, MA: Harvard University Press, 1997), 132.
59. Bertrand Gervais, "Reading as a close encounter of the third kind: An experiment with Gass's 'Order of insects,'" in Brown, *Insect Poetics*.
60. Joanne Elizabeth Lauck, *The Voice of the Infinite in the Small: Re-visioning the Insect-Human Connection* (Mill Spring, NC: Swan, Raven, 1998).
61. Quoted in Copeland, *Cockroach*, 168.
62. Jeffrey A. Lockwood, "The moral standing of insects and the ethics of extinction," *Florida Entomologist* 70 (1987): 70–89; Jeffrey A. Lockwood, "Competing values and moral imperatives: An overview of ethical issues in biological control," *Agriculture and Human Values* 14 (1997): 205–10; Jeffrey A. Lockwood and Alexandre V. Latchininsky, "Philosophical justifications for the extirpation of non-indigenous species: The case of the grasshopper *Schistocerca nitens* (Orthoptera) on the Island of Nihoa, Hawaii," *Journal of Insect Conservation* 12 (2008): 235–51.
63. Jeffrey A. Lockwood, "Not to harm a fly: Our ethical obligations to insects," *Between the Species* 4 (1988): 204–11.
64. Bill Devall and George Sessions, *Deep Ecology: Living As If Nature Mattered* (Layton, UT: Gibbs Smith, 2001); Nel Noddings, *Caring: A Feminine Approach to Ethics and Moral Education*, 2nd ed. (Berkeley: University of California Press, 2003); Wilson, *Biophilia*.
65. Jeffrey A. Lockwood and Scott P. Schell, "Decreasing economic and environmental costs through reduced area and agent insecticide treatments (RAATs) for the control of rangeland grasshoppers: Empirical results and their implications for pest management," *Journal of Orthoptera Research* 6 (1997): 19–32; Jeffrey A. Lockwood, Scott P. Schell, R. Nelson Foster, Chris Reuter, and Tahar Rachadi, "Reduced agent-area treatments (RAAT) for management of rangeland grasshoppers: Efficacy and economics under operational conditions," *International Journal of Pest Management* 46 (2000): 29–42.
66. Kellert, "The biological basis for human values of nature," 42.
67. Erika Mae Olbricht, "Made without hands: The representation of labor in early modern silkworm and beekeeping manuals," in Brown, *Insect Poetics*.

68. Jeffrey A. Lockwood, *Six-Legged Soldiers: The Use of Insects as Weapons of War* (New York: Oxford University Press, 2008).
69. Mike Quinn, comp., "Giant spider web in an East Texas State Park—2007," Texas Entomology, http://texasento.net/Social_Spider.htm (accessed July 25, 2012).
70. David N. Byrne, Edwin H. Carpenter, Ellen M. Thoms, and Susanne T. Cotty, "Public attitudes toward urban arthropods," *Bulletin of the Entomological Society of America* 30 (1984): 40–44.
71. Kellert, *Kinship to Mastery*.
72. Wilson, *Biophilia*, 12.
73. Ibid., 96.

CHAPTER 10

Entomapatheia: Can't We Just Live and Let Live?

Physicists aspire to a "theory of everything" as a culmination of their science. And now it seems that such an overarching framework has been discovered in psychology, thanks to the explanatory power of evolutionary biology. However, Newtonian mechanics was once seen as a complete account of the physical world. Sometimes a wonderfully complete theory, like beautiful woodwork, relies on a veneer that is elegantly polished but prone to splitting when stressed. And a theory can be bent by empirical observation and twisted by conceptual criticism.

A good theory aligns with the data. So the framework of biophilia should account for not only anecdotal evidence but also the findings of systematic studies. Stephen Kellert conducted the definitive study of our perceptions of invertebrates, with insects being the centerpiece of the spineless world.[1] As for biophilia, a negativistic attitude topped the list. Unsurprisingly, people don't like insects (along with crabs, spiders, scorpions, and ticks). Although the dominionistic view was not isolated in this study, the responses suggest that this attitude would have been near the top. So far, so good.

Rather unexpectedly, the next most common response was symbolic/aesthetic (combined in Kellert's analysis). We might guess that people mostly perceive insects to be compelling representations of powerful ideas and feelings rather than sources of beauty.

Then came a set of biophilic perceptions that one might be able to rationalize with a bit of creativity: utilitarian (most everyone knows that bees are pollinators, for example), ecologistic (perhaps insects are understood as contributing to the "balance of nature"), naturalistic (butterfly collecting is a quaint pastime), and scientistic (nature documentaries make insects interesting).

At the bottom of the list were moralistic and humanistic responses. People desired a world without cockroaches, fleas, mosquitoes, moths, and spiders.

Despite the Shorbs' tale of arachnid friendship, Eric Carle's stories about lovable crickets and ladybugs, and all of those butterfly nets and bug cages in toy stores, we just don't have warm feelings for insects and their multilegged kin.

All of this would make a more or less plausible, if not entirely convincing, account of entomophilia, except for one problem. Some scientists feel no affection for the theory of biophilia. And they have some valid, conceptual objections. But, as we shall see, there may yet be a way to understand the attentive—if not the infatuated—mind.

BIOPHILIA: WHAT'S NOT TO LOVE?

The idea that our deepest, most primal urge is to affiliate with all of nature, that not only our cars and detergents but even our psyche is Green, resonates with what we would like humans to be. But there are reasons to doubt that our inner lives are so environmentally virtuous.

How'd You Do That?

Harvey Lemelin, a professor at Lakehead University in Ontario, has spent a great deal of his career studying the relationships between humans and insects with regard to outdoor recreation and education. And his research has led him to become a prominent biophilia skeptic. His doubts start at the root of scientific inquiry—if one employs faulty methods, then the data aren't going to be convincing. As a social scientist Lemelin finds the approaches of the biophilia investigators seriously flawed: "[Kellert's] methodology is limited because it is for all intents and purposes Eurocentric, positivistic, and dualistic."[2]

Lemelin's methodological objections run still deeper. He is suspicious of tidy explanations of human behavior arising from simplistic schemes. Specifically, he finds employing evolutionary reductionism as the conceptual technique for understanding our responses to insects to be something like explaining the pyramids of Egypt solely with geometry. The physics of tetrahedrons is relevant, but it's hardly the whole story.

Evolution: Explanations or "Just-So Stories"?

The problem with any theory about prehistoric minds is that preliterate people didn't write down their thoughts. So we are free to construct all sorts of spiffy evolutionary tales without fear of refutation. For example, when pondering the origin of his ophidiophobia, Wilson concludes, "The direct and simple answer is that throughout the history of mankind a few kinds [of snakes] have been a major cause of sickness and death."[3] A major cause? A much stronger case could be made for mosquitoes, and we find these insects annoying but hardly frightening. When imagination substitutes for data, scientific explanations of how humans developed entomophilia can resemble Rudyard Kipling's tale of how the leopard got his spots.

Even if biophilic tendencies had evolutionary payoffs, there's a problem in making the leap to these proclivities being genetically encoded. Michael Soulé, a well-known evolutionary and conservation biologist, notes that the evolutionary fitness of a trait is often *inversely* related to its heritability:

> The human trait with highest estimated heritability (95 percent) is the number and pattern of fingerprint ridges, for example, a seemingly trivial characteristic from the standpoint of survival and reproduction. Yet the so-called fitness characteristics often have heritabilities below 30 percent. The number of heads in humans has a heritability of close to zero.[4]

Accurate predictions validate scientific theories, and it seems that biophilia fares rather poorly in this regard. Using the principles of biophilia, Soulé hypothesized that the human aversion to spiders would not show gender differences but would exhibit ethnic and cultural differences.[5] Oops. Contrary to his predictions, women are far more likely than men to manifest arachnophobia, and this fear is widespread among human societies.

Soulé also proposed that biophilia would lead one to expect little variation among mammals with regard to the ancient senses of smell and touch.[6] I grew up with dogs, and their responses to feces and carcasses would suggest that humans' and dogs' views of stinky, squishy objects are strikingly divergent. The prediction that humans and our furry kin would share primal perceptions is further undermined by the finding that disgust (being richly informed by tactile and olfactory cues) is unique to humans.

The most incisive critique of the evolutionary benefits of biophilia comes from a pair of South African scientists—an ecologist and an entomologist. John Simaika and Michael Samways argue that the environment takes care of itself without our meddling:

> There is no need for an innate affection for living things or nature at large. In other words, there is no need for a genetic basis for biophilia because there was never any need to evolve biophilic behavior.[7]

That is, loving nature had no adaptive benefit because the "bio" doesn't need our "philia." Other species should cringe when people show up and announce, "We're from humanity and we're here to help."

Like Lemelin, Simaika and Samways are deeply concerned about biodiversity, but they contend that the need to advocate for conservation stems from the fact that humans must *learn* to have an affinity for the natural world. Such empirical evidence gives rise to the next major objection to biophilia.

Biophilia Meets the Real World

If absence makes the heart grow fonder, humans should be head over heels in love with nature. With as many as twenty thousand species going extinct each year, we ought to be giddy about biodiversity. Of course, nearly all of these losses are our doing. So if we love nature, we sure have a strange way of showing it.

Kellert contends that "our material, emotional, intellectual, and spiritual well-being demands living abundance and variety."[8] However, city dwellers don't seem to be suffering—at least any more than rural folks who have lots of species around. Studies in the United States, Germany, and Japan reveal lukewarm support for maintaining biodiversity.[9] So what's up with humanity?

Our failure to care about other species could reflect ignorance about what we need to be healthy and happy. Cutting down tropical forests might be like eating donuts—satisfying in the short term but bad for us in the long haul. But this suggests that biophilia, if it exists, is a feeble inclination. For the most part, modern humans are too busy making a living for living things to matter much to us.

If we use the ability to name organisms as a measure of biophilia (if you can't remember someone's name, then there isn't much of a relationship), we are empirically indifferent to nature. A colleague and I taught a nonmajor science course called "The Biodiversity Crisis." We asked hundreds of students across several years to write down the common names of three native vertebrates, plants, and insects. These were mostly Wyoming kids, so they'd probably spent more time in the outdoors than the average American college student. Even so, fewer than half could name three native vertebrates, about a quarter could name three plants, and less than 5 percent could name three insects.

A clever theorist might be able to explain how ignorance is actually a form of biophilia, but that takes us to the next objection.

Biophilia Explains Everything—and Nothing

Critics point out that Wilson's foundational definition of biophilia, "the innate tendency to focus on life and lifelike processes,"[10] doesn't exclude much of anything that people love. Any and all organisms count, from bacteria to begonias to bison. So do habitats and whole landscapes, from ponds to prairies. For that matter, we also stare dreamily into the night sky, and there could be life out there, after all.

Perhaps more worrisome for biophilia advocates is our demonstrable affinity for nonliving objects. The Indian ecologist Madhav Gadgil points out that Hindus worship objects that are vital to their survival. He contends, "As artifacts have come to play an increasingly significant role for human societies, they have usurped the place that natural elements and living organisms once played as objects of human reverence."[11] And one might suppose that such affinities extend back to stone tools. Churches replaced sacred groves, and tractors replaced draft animals. Gadgil's conclusion is that humans are fascinated by any complex entity.

Biophilia advocates might point out that these inanimate objects have "lifelike" qualities. Machines respond, move, and self-regulate. Computer software can have a "bug" or be infected with a virus, and skyscrapers resemble termite mounds. However, virtually everything has *some* lifelike feature, so how could any expression of love not be a manifestation of biophilia?

The problem of overreaching is even more serious than the pervasiveness of organic features in the world. Given an ant crossing the kitchen counter, it seems that no matter what our response, we're engaged in biophilia. Perhaps we think, "What a marvelous little creature!" Such a positive reaction qualifies as entomophilia by all accounts. Or maybe we think, "Damn that little home invader!"—and this represents a dominionistic or negativistic version of entomophilia.

When I was in junior high, it would have been keen if no matter how a girl responded, it was evidence of affection. A dismissive sneer would have been as affirming as a warm smile. But what works for an adolescent isn't viable for a scientist. If every emotional response to essentially every object, place, and process counts as biophilia, then we have a hypothesis that cannot be falsified.

If biophilia is logically flawed, this means trouble. After all, the core of the theory depends on rationality, which brings us to the final objection.

The Nature of Entomophilia—and Emotion

Biophilia's founders tend to scientism, the view that everything is explicable in terms of science. E. O. Wilson contends, "If the mind is the creation of the brain, then it must be subject to material explanation."[12] For critics, that's a big "if" and a huge leap of faith regarding the power of reductionism.

Harvey Lemelin is not among the faithful. He sees humans as profoundly complicated social creatures. Our contradictory behaviors and beliefs defy rationalistic accounts. He points to the irrationality of our food choices, in which we deem repulsive the consumption of arthropods fed on fresh grass (say, grasshoppers), while we pay top dollar in fine restaurants to dine on arthropods fed on sewage and decay (say, lobsters).[13]

But Wilson and Lemelin have one thing in common—they're both deeply committed to insect conservation. They might even agree that when it comes to fostering entomophilia, the reasoned arguments of science aren't working. As Michael Soulé puts it, "Facts compute, but they don't convert."[14] Lemelin takes the process of cultivating genuine connections to insects one step further.

While working with the Mohawks, Lemelin came to understand their world of animism, in which all things are alive and worthy of respect. When one perceives a kind of life force connecting all of existence, then "it is not surprising that insects are valued."[15] Exactly what role cockroaches, mosquitoes, and hornets have is not a simple matter even for the animist, but a Mohawk elder explained to Lemelin that insects have their place, "even if we can't figure it out."[16]

Biophiliacs might try to explain how the Mohawks' worldview was adaptive for these native people, but this is surely an insufficient account of spirituality. And while Wilson preaches the gospel of entomophilia, Lemelin would be happy to convert others to something less psychologically radical: ambivalence.

BIO-AMBIVALENCE: DON'T ASK, DON'T TELL

Biophilia as a descriptive theory of the human psyche has problems, so what about its potential as a prescriptive theory? That is, if we aren't actually biophilic perhaps we ought to be. Advocates have pursued two lines of argument in this regard.

First, there could be a moral obligation to become biophilic. David Orr contends that "for every biophobe others have to do that much more of the work of preserving, caring for, and loving the nature that supports biophobes and biophiliacs alike."[17] In short, biophobes are contemptible freeloaders, the equivalent of sociopaths. However, few (if any) people are biophobic, fearing or loathing all life. Rather, they don't like some living things. We don't expect everyone to love everybody; should we expect everyone to love maggots and cockroaches?

The second view is that we ought to be biophilic for our own mental health. Kellert asks us to a imagine "a beautiful and peaceful world, where the horizon is rimmed by snowy peaks reaching into a perfect sky . . . except for one thing—it contains no life whatsoever. . . . This is a world where people would find their sanity at risk."[18] I'm dubious. Maybe people would be happy with having just libraries and museums. Nor is there anything more than correlations and anecdotes to support the more extreme claim that "limited opportunities to express biophilia may lead to psychopathologies."[19] Lots of sane people don't watch birds or collect butterflies, and there are plenty of psychologically imbalanced nature lovers.

These lines of argument (moral and mental) presume a dichotomous framework: sane philia or demented phobia. But there is a third option. Gays in the military are like insects in society. Well, not exactly, but there is an interesting parallel. Even those who find homosexuality repulsive often concede that we should treat other people decently. This might not mean that homophobes are required to become homophiles, but at least they should be homo-ambivalent. Maybe the same goes for insects.

* * *

Between entomophobia and entomophilia is entomapatheia (actually I made this term up with the help of a colleague versed in Greek; *apatheia* refers to the state of not being affected by something or of lacking strong feeling). We could just live and let live. In fact, this is what Harvey Lemelin proposes. When I asked him what would be the ideal emotional response to insects, he replied, "That it is okay to see insects as awful and awesome, that it's also okay to be ambivalent." Being a social scientist, he's willing to accept that "humanity, like nature, is diverse. Polymorphic interpretations and tendencies are natural. It's just the way we are."[20] Lemelin is not okay with killing insects because we are simply ignorant or unjustifiably fearful, but when it comes to the psychology of conservation he maintains that we should simply "strive for tolerance"—not infatuation.

This might seem to be a less than idealistic position for an activist, but Lemelin knows whereof he speaks. He admits to being a recovering entomophobe. For years he was terrified of insects and killed them on sight. This fear-driven

Figure 10.1
Parents and teachers can play an important role in cultivating children's perceptions of insects. The girl in this photograph appears apprehensive—exhibiting neither strong affection nor antipathy toward the insect. Developing a benign indifference toward insects (entomapatheia) may be all we can reasonably expect, and this can constitute a psychologically healthy state (image from Public-Domain-Image.com).

destruction was reinforced by his upbringing, and not until he began working with the Mohawks as an adult did he learn to accept insects. He still kills them when "absolutely necessary," but such conflicts are now rare. In this regard, entomapatheia does not mean a mindless indifference that would lead to one's apartment being overrun with bed bugs or one's crops being devastated by locusts. Rather, it entails allowing insects to go about their lives and minding our own business in the vast majority of cases in which there is no cause for conflict. For Lemelin, mastering this outwardly simple and inwardly challenging approach required years of intense practice, listening to teachers, reading books, and engaging insects.

Given his journey, Lemelin finds the goal of infusing entomophilia in Western society a bit like seeking world peace. It's good to have a lofty aspiration guiding one's work, but we're not all cut out to collect butterflies, watch birds, paint wildflowers, dance in meadows, or frolic in forests. As Lemelin sees it, "If some people prefer to not want to know about or interact with insects, that's fine, but they should at least be tolerant of the 'other.'"[21] And some of the staunch biophiliacs might concede his point.

Michael Soulé has suggested that the average person's struggle for material and emotional survival moves biophilia to the psychological back burner.[22] Even Stephen Kellert seems to allow that biophilia provides higher-order fulfillment in a world where basic requirements are often unmet.[23] Perhaps biophilia is a psychological luxury, a lofty step on our hierarchy of needs.

E. O. Wilson admits that while it is possible to change one's response to a reviled creature, "the adaptation takes a special effort and is usually a little forced and self-conscious."[24] That's quite a lot to ask out of someone who is wondering where she's going to get medical care or where he's going to find a job. And David Orr dismounts from his moral high horse to recognize that not everyone who fails the acid test of biophilia is a social leech (these creatures being even more despicable than insects). Although he believes that "life ought to excite our passion, not our indifference," he is willing to settle for people having a "respectful regard" toward other beings—a position that seems close to Lemelin's notion of tolerance.[25]

Before we conclude that evolutionary psychologists, conservation biologists, social scientists, and environmental activists are singing "Kumbaya," I'd like to propose a way of understanding our relationship to insects that gives evolution its due while allowing culture its place in human psychology.

INSECTS: NATURE'S VALENTINE CARDS AND STOP SIGNS

The famed mother-son team of Lynn Margulis and Dorion Sagan (brilliant biologist and acclaimed writer) see the taproot of biophilia extending into the earliest life forms: "Perhaps it would be better to speak of prototaxis—the generalized tendency of cells and organisms to react to each other in distinct ways. . . . Let us think then of both positive and negative biophilia (sometimes called biophobia) as aspects of global prototaxis."[26] This might be a bit too primal to account for the human psyche, but other scientists also see a parallel: "Children who throw stones at birds and children who feed birds are both responding to what may be an innate tendency to focus their attention on living things."[27] In psychological terms, the feathered target is salient.

The gurus of biophilia tacitly endorse salience as the psychological foundation for the emotions that follow. Kellert contends that connections to living beings arise from "our instinctive inclination to react strongly to certain elements of nature."[28] Wilson—a romantic at heart, his scientism notwithstanding—puts it this way: "The brain appears to have kept its old capacities, its channeled quickness. We stay alert and alive in the vanished forests of the world."[29]

Although psychiatrist Aaron Katcher and psychologist Gregory Wilkins are best known for their work on pet therapy, they have weighed in on the broader context of biophilia—and their view aligns with the idea that both evolution and culture influence our responses to other creatures:

> If biophilia exists, then it most probably exists as a disposition to attend to the form and motion of living things and, for animals at least, incorporate them into the social environment.[30]

So what if evolution has not predisposed us to fear or love insects, but we're simply primed to pay attention to them? Perhaps insects are to life forms as red is to colors. Red draws our eye to important objects. Many snakes and insects use red as a warning color for would-be assailants, while many flowers and fruits use red to attract pollinators and seed dispersers with the promise of a sweet snack. And society uses red to draw attention in both positive (e.g., commercial packaging) and negative (e.g., warning lights) ways.

Here's the entomological concept in a nutshell (which is where you'd find a pecan weevil larva, by the way). Insects and their kin have been important in human ecology—sometimes as dangers, sometimes as resources. Evolution assures that we notice these creatures, while culture shapes our responses to these multilegged valentine cards and stop signs.

Let's consider a three-step process in light of what we know about humans and insects.

Step 1: Your Attention, Please!

When we combine insects and spiders that can immediately harm us by biting or stinging with those that can directly benefit us by being edible or inspiring, we end up with a very long list of creatures. And then if we throw in all of the indirect costs (such as spreading germs and eating our crops) and benefits (such as pollination and preying on pests), we're looking at hundreds or thousands of species. Sure, there have been insects that don't much matter to our well-being, but there are also red things that don't pertain to our survival, like sunsets and sandstone. The point is that small flitting or crawling objects were often worth our attention.

Step 2: Now That I Have Your Attention . . .

When a kid in a New Guinea village sees a dinner-plate-sized spider, it means one of two things—that thing could deliver a painful bite or that thing would be scrumptious on a dinner plate. These indigenous children are embedded in a culture that teaches them how to distinguish between these two kinds of spiders. But in the past ten thousand years most societies have distanced themselves from wild dangers and foods. According to Jared Diamond, spiders and insects are no longer worth the cultural investment of teaching kids to differentiate species.[31] We simply generalize the initial focusing of our attention into an aversion, which mostly works in the modern world.

Step 3: Open Eyes and Tolerant Minds

The first two steps of the evolutionary-cultural response to insects constitute a twist on the "prepared learning" hypothesis, which suggests that humans are predisposed to learn a fear of insects and spiders. This hypothesis is both too weak (we aren't just ready to learn, we're highly tuned to attend) and too strong (we're primed to learn about the things that grab our attention, but we're as ready to absorb positive judgments as we are negative ones).

We can't help but see insects, but we can decide what to do next. And this takes us to the third step. In recent centuries we've abhorred insects, and today's environmentalists want us to love them, but perhaps Lemelin has it right. When a child says, "Look, there's a bug!" maybe a few parents can respond, "Well dear, bugs have piercing-sucking mouthparts. That's actually a ladybird beetle—a beautiful and gentle creature that symbolizes the virtues of the Virgin Mary and helps our garden by eating aphids." But in the real world, we might do well with, "That's right, honey. And if you leave it alone, it will leave you alone."

Maybe entomapatheia is not just a workable approach to other species but also provides a foundation for our interactions with one another. Insects provide us with opportunities to practice tolerance of differences. One of the great challenges of the twenty-first century is how we will respond to alien religions, cultures, genders, politics, and species. Perhaps this is overextending the power of cultivating an acceptance of insects, but why can't this be a way to begin overcoming our xenophobia—fear of the unfamiliar? Gay men kissing, women in burkas, people speaking Spanish, and spiders in the basement. Who knows, if kids learn that they can coexist with radically different creatures, maybe they'll figure out that they can live among not-like-us humans.

NOTES

1. Stephen R. Kellert, "Values and perceptions of invertebrates," *Conservation Biology* 7 (1993): 845–55.
2. Author e-mail interview with Raynald Harvey Lemelin, June 21 and July 1, 2011.
3. Edward O. Wilson, *Biophilia* (Cambridge, MA: Harvard University Press, 1984), 96.
4. Michael E. Soulé, "Biophilia: Unanswered questions," in *The Biophilia Hypothesis*, ed. Stephen R. Kellert and Edward O. Wilson (Washington, DC: Island Press, 1993), 448.
5. Ibid., 443.
6. Ibid., 445.
7. John P. Simaika and Michael J. Samways, "Biophilia as a universal ethic for conserving biodiversity," *Conservation Biology* 24 (2010): 903–6.
8. Stephen R. Kellert, *Kinship to Mastery: Biophilia in Human Evolution and Development* (Washington, DC: Island Press, 1997), 187.
9. Ibid., 188.
10. Wilson, *Biophilia*, 1.
11. Madhav Gadgil, "Of life and artifacts," in Kellert and Wilson, *The Biophilia Hypothesis*, 372.
12. Wilson, *Biophilia*, 47.
13. Lemelin, author interview.
14. Michael Soulé, "Facts compute, but they don't convert," interview by Lisa Jones, *Sierra*, July/August 2003, http://sierraclub.org/sierra/200307/interview.asp (accessed July 25, 2012).
15. Lemelin, author interview.
16. Ibid.
17. David W. Orr, "Love it or lose it: The coming biophilia revolution," in Keller and Wilson, *The Biophilia Hypothesis*, 419.
18. Kellert, *Kinship to Mastery*, 172.
19. Eleonora Gullone, "The biophilia hypothesis and life in the 21st century: Increasing mental health or increasing pathology?," *Journal of Happiness Studies* 1 (2000): 293–321.
20. Lemelin, author interview.
21. Ibid.
22. Soulé, "Unanswered questions," 453.
23. Kellert, *Kinship to Mastery*, 9.
24. Wilson, *Biophilia*, 95.
25. Orr, "Love it or lose it," 434.
26. Dorion Sagan and Lynn Margulis, "God, Gaia, and biophilia," in Kellert and Wilson, *The Biophilia Hypothesis*, 347.

27. Aaron Katcher and Gregory Wilkins, "Dialogue with animals: Its nature and culture," in Kellert and Wilson, *The Biophilia Hypothesis*, 175.
28. Kellert, *Kinship to Mastery*, 148.
29. Wilson, *Biophilia*, 101.
30. Katcher and Wilkins, "Dialogue with animals," 193.
31. Jared Diamond, "New Guineans and their natural world," in Kellert and Wilson, *The Biophilia Hypothesis*.

CHAPTER 11

Back to the "Real" World: Good Night, Sleep Tight . . . or Maybe Not

I have a colleague in sociology who tells his introductory students at the beginning of the semester, "Each of you is a special and unique individual." This reassures them of what they've been told for years. But then he adds, "Just like the other seven billion people on the planet." And so their sense of individuality is starkly challenged. My colleague is a darkly humorous fellow, but then humans are funny creatures. On one hand, we aspire to be normal. We want to fit in physically and psychologically—not too fat or skinny, not too aggressive or passive. On the other hand, we want to be unique. We'd like to be in the upper tail of the distribution when it comes to mental qualities such as intelligence (aren't you confident that you are above average?) or empathy (don't you think that you're more compassionate than the typical American?).

When it comes to our responses to insects, I suspect that most of us want to be somewhere near the middle of the curve. Perhaps we imagine that we would exhibit greater courage than our compatriots when confronting a monstrous ant in the desert or a plate of sautéed water bugs, but in daily living we aspire to normalcy. That said, balance allows for experiences at the margins. A normal mother reads her daughter the enchanting story of *Charlotte's Web* and grabs a can of Raid to dispatch an earwig in the kitchen. And a normal dad plays the video of *A Bug's Life* for his son and mashes a house spider underfoot. Most of us figure that the weirdoes are those stuck at the extremes. We're not like the entomophiles who harbor spiders as housemates or derive sexual satisfaction from the tickling of ants. And we're certainly not crazy entomophobes who run panic-stricken into the streets upon encountering a cockroach or douse themselves with chemicals to kill imaginary lice. We figure that we're perfectly functional and far from the deep end of irrational fear. But what really stands between our presumed sanity and an explosion of debilitating terror?

Is our apparent normalcy a matter of our internal life or our external conditions? It's rather easy to imagine that we're rational about insects when the modern world provides an environment that is largely devoid of these creatures. We might have to stuff the odd roach motel under the sink, apply some mosquito repellent, or spray the roses for aphids, but better living through chemistry means that insects are under control—along with our fears. And when these creatures threaten our well-being, we can rely on professionals to sanitize our homes and yards.

However, what if the insects refuse to submit? What if our lives and bodies are under siege? What causes apparently normal people—the ones who are like us—to destroy their furniture, burn their clothes, and abandon their homes?

What does it take to produce an infested mind? The answer is a quarter-inch-long bug.

GOOD NIGHT, SLEEP TIGHT . . .

At the beginning of the twenty-first century, a crack appeared in the facade of our insect-free society. From behind headboards and beneath mattresses, bed bugs crept into the national consciousness.[1] The reason for the upsurge appears to have been a combination of our banning formerly effective insecticides and the insects evolving resistance to our chemicals. The suddenness of the resurgence is probably an illusion. Having virtually extirpated bed bugs for a half century, entomologists, dermatologists, and public health officials often overlooked infestations and misdiagnosed patients until the insects reached epidemic levels.[2]

From 1999 to 2000, there was a 300 percent increase in the number of bed bug calls to the Orkin pest control company,[3] and since 2001, the National Pest Control Association reports, complaints about bed bugs have increased up to a thousandfold.[4] Florida saw a tenfold increase in bed bug service calls in just two years,[5] and New York's Health Department tallied 4,084 bed bug violations in 2010, compared to just 82 in 2004.[6]

Although the East Coast has drawn the greatest attention, by now every state has bed bugs. And it's a good bet that most every major city in the world supports a flourishing population. A recent news article declared:

> Tiny, bloodthirsty and relentless, the insect pest that became a nightmare for millions of New Yorkers has found its way to Paris, spreading fear of an outbreak that could turn into the ultimate tourism-killer.[7]

The dire prognosis is warranted given that 20 percent of Americans say they have changed their travel plans for fear of bed bugs.[8] Cold, dry places with

Figure 11.1
Bed bugs are not terribly frightening in terms of their size (about that of an apple seed) or by virtue of their harm (bites are irritating but the insects don't transmit pathogens). Rather, it is their behavior that so deeply disturbs people. Bed bugs emerge at night to feed on our blood and then disappear during the day (image by Piotr Naskrecki through Creative Commons and the Centers for Disease Control and Prevention).

few people have less of a problem, but the housing staff at the University of Wyoming—where I'm on the faculty and can attest to the cold and dryness—find bed bug infestations every year in the dorms. There aren't a hundred thousand of the beasts as in some New York apartments, but it doesn't take many of these insects to change lives and minds.

* * *

At night, bed bugs crawl out from their hiding places to feed on blood. As unappealing as this might be, they do not transmit pathogens and their bites are relatively benign (some individuals have severe reactions, but most simply develop small red welts that itch like mosquito bites). However, the thought of having our bodies invaded while we sleep drives people to extreme measures.

A New York family reported throwing out forty garbage bags filled with clothes and toys and paying $3,500 to a pest control company after a bug-sniffing dog (a popular method of detection—and deception) indicated an infestation.[9] These people got off cheap. Another family claims to have spent more than $70,000 in the course of their entomological odyssey.[10] On the West Coast, a Los Angeles woman destroyed half of her clothing and most of her furniture to assure that the insects were gone.[11] While the financial costs can be staggering, the most devastating damage done by bed bugs is psychological.

A bed bug infestation can trigger extraordinary behaviors. An otherwise normal university student revealed:

> Within a week, it got to a point where I was sleeping on top of blankets in the middle of my living room floor surrounded by a tape barricade. I'd wake up in the middle of the night and freak out because my blanket crossed the tape, making a potential bridge for invaders—and this seemed completely normal to me.[12]

The Internet is filled with accounts of the incredible lengths to which people will go in trying to keep the bugs at bay—including the contemplation of suicide.[13] There are now effective, if expensive and intrusive, ways of eliminating these insects. But, as that college student cogently recognized, "While bed bugs infest your apartment, they also take up residence in your mind."[14]

Months after the insects have been exterminated, people struggle with anxiety, interpreting every tickly sensation as evidence that the bed bugs have returned. Some cases take on the features of delusory parasitosis: "You don't feel safe climbing into bed to sleep. Every little itch you wonder, is that a bedbug? You think they're there all the time."[15] Insightful parents recognize that a bed bug infestation can mold a child's future and fears. One mother described her efforts to desensitize her son after their home was debugged: "I had him watching animated movies like *Antz* and *A Bug's Life*. Anything that treats bugs in a friendly way. Maybe we'll go to a farm and let him actually catch bugs."[16]

Entomologists and exterminators are on the psychological front lines. An insect diagnostician at Cornell University has received bits and pieces of every imaginable kind of household detritus in the mail from people wanting to know if they have bed bugs. And the samples are often accompanied by worrisome notes:

> I am super paranoid that I have bedbugs. No bites, just crazy paranoia. This is all the evidence I have. [Sample] C looks like a cockroach, but I don't know about the others. Please give me a careful and thoughtful definitive answer.

And:

> [Found] a few weeks ago, not sure. Would like to know if they are bedbugs. I was seen in the E.R. approx. three weeks [ago], and was told I had insect reaction on my calfs [*sic*].[17]

A pest control operator who works exclusively on bed bug suppression recognizes that "most of my job is not removing the bed bugs. It's removing the psychological effects after they're gone."[18] But often, exterminating fear is best left to the mental health professionals.

Although there may not be any psychologists or psychiatrists specializing in mental disorders arising from bed bug encounters (real or imagined), at least one New York midtown psychotherapist who treats people suffering from obsessive-compulsive disorders has built a part of his practice around those seeking help with their infested minds.[19] To get a sense of the scope of the problem, I did an online search using the paired terms "bed bug" and "PTSD"; there were nearly two million hits.[20] Pairing "bed bug" with "anxiety" and "depression" yielded half a million hits each. This isn't sophisticated epidemiology, but the results do indicate that the problem appears to be enormous.

In the midst of this pandemonium it is easy to miss the psychological forest for the phobic trees. In short, why do bed bugs elicit such dramatic responses? While we've suppressed insects for decades, most of us have encountered blood-feeding mosquitoes and nocturnal cockroaches. A bed bug is more or less a chimera of these two insects, neither of which evokes abject horror in most people. And therein lies the key.

According to Susan Miller, "Horror is the likely response when little can be done to resist the invasion of some powerful outsider . . . a response to what truly is alien and other-than-self [that] threatens to substitute its being for our own."[21] And few creatures are as horrifying as vampires—except, perhaps, bed bugs. People battling infestations have made the connection: "They're having babies every day. They spawn vampires."[22] And a *Washington Post* reporter proposed that *Twilight* and *True Blood* added fuel to the bed bug fire, noting that "the lentil-sized beasties are generally nocturnal, prefer human flesh and bite us to suck our blood."[23]

SIX-LEGGED VAMPIRES: AN ENTOMOLOGICAL NIGHTMARE

Connecting our horror of bed bugs with our horror of vampires is not meant to make light of those who suffer the psychological trauma of an infestation. Indeed, quite the opposite. Jeffrey Weinstock has undertaken a fascinating scholarly study of why vampires are so persistent in human culture[24]—and I propose that these perennial qualities align with the biology of bed bugs to create a perfect psychological storm. My effort is to draw out these connections as a way of understanding, not dismissing, the damage done to the human psyche by bed bugs. Cinematic vampires, which are the focus of Weinstock's analysis, tap into deep, dark places of the mind, the sorts of recesses where one finds bed bugs.

Figure 11.2
The Austrian artist Ernst Stöhr captured the essence of our fears in his *Vampir* (1899). The creature depicted in this work evokes a sense of carnal power and unrestrained sexuality that subdues her helpless victim during the night. We are darkly fascinated by the primal power of this monstrously "other" being to infest our psyche and draw us into the realm of the undead (image through Wikimedia Commons).

Bugs in the Boudoir

According to Weinstock, "Vampire narratives are always about sex."[25] Vampires are driven by erotic desire to consume the life force of their victims. Bed bugs might be less lascivious, but their unwanted intimacy with our bodies is similarly appalling. Our beds are where we are sexually vulnerable, where we lie naked and make love—and so finding spots of our own blood on the sheets in the morning is profoundly disturbing. Bed bugs don't simply come into our beds, they often live within their crevices, continuously violating the most intimate and private place in our lives. Add to this the fact that insects have been interpreted as symbols of penetration,[26] and we have a psychosexual nightmare in the works. In fact, bed bug males mate via traumatic (or hypodermic) insemination, during which they pierce the body wall of the female with their genitalia and inject sperm into the abdominal cavity.

The sexualization of bed bugs is not merely fanciful. Interviews with regular folks reveal a profound connection to the realities of contemporary intimacy. Romances fail because of infestations: "Bed bugs are definitely a very creepy aspect to dating at this point."[27] The irrational extreme of people's reactions to these insects is revealed by the observation that, "for some, the prospect of bed bugs is even scarier than a sexually transmitted disease."[28] As one woman notes, "Well, a condom can't protect you from bed bugs, that, I think, is for sure."[29] Other diseases are also evidently preferable to bed bugs, as evidenced when one fellow's wife saw that he had a rash, contemplated the possibility of bed bugs, and said, "I hope to God you have shingles."[30] In extreme cases, sexually transmitted diseases are perhaps the least of the dire possibilities:

> I'm more terrified of bedbugs than I am of herpes, terrorist attacks, ebola and the black plague combined. Anytime I see a mattress on the sidewalk I pretty much cross the street to get as far away from it as possible out of fear that bedbugs [*sic*] will somehow jump off it onto my clothes.[31]

Not only can we use paper covers to protect us from the pathogens lurking on toilet seats, we now have "Seat Defenders"—trash bags modified into giant condoms to cover theater seats and prevent bed bugs from infesting us.[32] In the most extreme cases, people perceive these insects as six-legged rapists:

> There are few things in the world that can make you and your home feel as *violated* [emphasis added], broken, and completely, completely hopeless as a bedbug infestation. The different aspects of bedbugs . . . how much of a stigma they have and how unable people are to be frank about them [not unlike being a victim of sexual assault]—could equal a massive public-health crisis.[33]

A creature that lives in our bed, crawls under our sheets, makes contact with our naked skin, and penetrates our flesh is nearly impossible not to perceive in sexual terms—whether vampire or insect.

Fascinated by Bed Bugs

What makes for a gripping tale of horror is a monster that is far more interesting than the people who pursue it. The vampire represents hedonistic, physical excess, living solely to satisfy its own pleasure. The animalistic drive for feeding and reproduction is simultaneously frightening and alluring. In contrast, "Those who seek to destroy the vampire, the agents of cultural repression, cannot help but seem priggish and impotent."[34] Sound familiar?

Consider that the Illinois Bed Bug Task Force—a governmentally officious name—was created "to study the increase in bedbug infestations and make specific recommendations for public policy measures to combat this growing public nuisance."[35] One might imagine a gathering of middle-level managers, narrowly trained technocrats, and pocket-protected scientists armed with spreadsheets while giving one another pedantic lectures about the arcane importance of fruitless regulations. Be honest, would you rather read about nocturnal, blood-feeding insects or the lives of bureaucrats, nerds, and killjoys? Frankly, entomology is interesting because insects are strange and scary, not because insect scientists are charming and dashing (of course, some are).

The battle against bed bugs is not being waged by the CIA, emergency room physicians, or professional athletes. The pursuers of these six-legged vampires are not subjects of reality television programs (except perhaps for Billy the Exterminator, but he's nothing without a cast of charismatic pests). And for the most part, they're impotent. In fact, the bugs have us battling each other while they dine on our blood—and our insect-catalyzed conflicts are probably the most intriguing part of the human pursuit. The general manager of the Chancellor Hotel on Union Square in San Francisco has dedicated a full-time position to a bed bug inspector in an effort to avoid having his establishment blacklisted by travel advisers.[36]

When prevention fails, our internecine tiffs take on an even nastier tone, although the bloodletting is still only metaphorical. Because the bed bugs exist outside our legal system, we resort to attacking each other. In their efforts to appear as if they are in control, people sue one another to attribute blame while the actual cause of the damage hides under the box springs. Most bed bug suits are settled out of court for less than $5,000,[37] but there has been one judgment in which a brother and sister were awarded $382,000 after they were accosted by bed bugs in a Motel 6.[38] Maybe they should've left the light on.

The Return of the Bed Bugs

Weinstock tells us that even when it seems that the vampire is destroyed, some trace or essence persists, and "all that's needed to revive him is a little blood, a voodoo ritual, or the removal of a cross from the corpse."[39] The entomological trace could be bloodstains on the bed sheets or the sickly sweet odor that some people say lingers after bed bugs feed. The insects can go for months without food and then be revived by a little blood. And when Weinstock notes that each time a new movie portrays Dracula, "he embodies a new structure of feeling, a different awareness, an altered set of fears and desires,"[40] one has to think that this time Americans are awakening to a modern villain from which even wealth and hygiene can't protect us.

Bed bugs are an age-old problem, probably beginning as soon as humans consistently bedded down in shelters. With the advent of DDT, we thought our nemesis had been defeated. A generation of physicians had passed over the textbook chapter on these insects and had all but forgotten the diagnostic signs of an infestation (how many doctors today would recognize the early symptoms of yellow fever or polio?). But as a writer for the *New Yorker* put it, "Man plans. Bedbugs laugh."[41] Those who have had bed bugs aren't laughing;

they're waiting. A great deal of their anxiety comes from the sense that the insects aren't really gone, that they are biding their time, "lurking in the shadows for the first opportunity to feed."[42]

Web-based chat sites are filled with speculation regarding how long bed bugs live (in reality, two to eighteen months) and, more disturbingly, how long they can lie in wait. Although bed bugs prefer a weekly meal, it is clear that they can live a long time without food. Some panicky posters assert that the insects can lie dormant for a year and a half, others insist on eight weeks. As one person put it, "The worst thing about bed bugs is that they live forever—up to a year without feeding!"[43] Whatever their actual life expectancy with regular engorgement or between meals, it's clear that they live far longer than we find assuring. Like vampires, they seem (nearly) immortal and destined to return.

Bugs versus Technology

To understand the intimate conceptual linkage between vampires and bed bugs, consider this description and simply substitute the latter for the former:

> The vampire is . . . defined in relation to and, in many cases, produced by particular technologies of detection, determination, and destruction. Vampire narratives thereby function as referendums on the inadequacies, perils, and promises of modern science and technology.[44]

Bed bugs remind us that our chemical cleverness, hygiene standards, and building codes are no match for the evolutionary patience of insects. The first response of many people to an infestation is to spray insecticides and set off foggers, both of which are utterly inadequate and ultimately perilous despite cultural assurances that we have the power to "kill bugs dead."

With the failure of today's remedies, victims often either turn to futuristic tactics generated by a cottage industry of entomological entrepreneurs or pine for the solutions (and presumably not the associated problems) of yesteryear—DDT, propoxur, and their ilk. But promises that new products will kill the bugs on contact "by breaking down the enzymes" and hopes that old products will work again despite insecticide resistance serve to spotlight our hubris (and gullibility).[45]

When a vampire afflicts a community in the movies, the hysterical populace relies on an expert to identify the source of their suffering and to provide a means to defeat the creature, a narrative that is echoed today by homeowners, apartment supers, and hotel managers. The storied weapon is usually a kind of theological pesticide (crucifix, holy water, communion wafer) or scientific device (inoculation or light grenade).[46] Today, the most viable technology to defeat insectan bloodsuckers appears to be heat. Superheated air (and steam applied to crevices in some cases) creates a household fever that cooks the bugs. The simplicity of this approach aligns with society's growing distrust of ever more scientifically sophisticated and technologically complex solutions to our problems.

But, of course, we also rely on the next generation of technology for protection—the holy water of social media. A traveler can search various websites to find places where others have reported bed bugs. There's even a smartphone app that allows users to see if prospective lodgings are infested and to report new encounters. There is, however, no vetting of these claims, and it is easy to believe that just as panicked townspeople falsely accused their neighbors of being vampires (or witches), nervous tourists are likely to see every speck as a potential bed bug and every pimple as a possible bite. Once again, the vampire reveals the promise and peril of technology.

Bed Bug as "Other"

According to Weinstock, the vampire is a distillation of whatever a society considers alien. He asserts that the creature "condenses a constellation of culturally specific anxieties and desires into one super-saturated form."[47] For example, *Dracula* was about (among other things) how the culturally inferior "others" from the backwaters of Central Europe infiltrated the heart of the empire. And insects are ideal psychological frameworks on which to erect modern us-versus-them distinctions.

Decent folks are bathed and deodorized, and their homes are clean and sanitized. Those lacking such hygiene are prone to bodily and household vermin. Like lice and cockroaches, bed bugs are associated with filth, which accounts for the social trauma they inflict. However, there is no actual connection between bed bugs and cleanliness. In fact, the misguided stigma associated with these insects tends to foster more severe infestations in wealthier and cleaner homes because of people's reluctance to have their neighbors see an exterminator coming and going. As a consequence, treatments are delayed until the pests reach intolerable numbers.[48]

To have bed bugs is to risk social isolation. The afflicted individuals are shunned, becoming "other" along with the vermin. Consider the reaction of a New Yorker upon learning that her immaculate home was infested:

> [She] was horrified, disgusted, and not a little concerned for her family. And although she is no snob, Margaret couldn't repress an uncomfortable thought: that people who live in multimillion-dollar apartments in the tonier precincts of the Upper East Side are just not supposed to have bedbugs.[49]

While Margaret feared for her family's health, she also worried about what the infestation meant for their social standing—and with good reason. People needing a place to stay during exterminations often find that their friends refuse to offer accommodations. Once the word is out, dinner invitations, play dates, and other social engagements are commonly canceled. Likewise, business associations can collapse if clients suspect the truth, which is why one caterer donned long sleeves and long pants during the hottest days of summer to conceal bed bug bites.[50]

Figure 11.3
As with vampires, zombies are a part of many cultures and exhibit some of the terrifying qualities of insects, particularly bed bugs. As portrayed in the 1968 film *Night of the Living Dead*, the monsters are very difficult to kill. Like insects, they lack consciousness and emotion while being keenly aware of external stimuli in their unwavering compulsion to feed on the flesh of living humans (image through Wikimedia Commons).

Susan Miller equates vampire victims with lepers, who "occupied the horrible station between life and death, the zone of the undead or walking dead . . . socially regarded as dead."[51] There may be no better account of the stigma that occurs when family, friends, and neighbors learn that someone is under the dark spell of bed bugs. For some people, it is like living in a real-world remake of *Invasion of the Body Snatchers* (a kind of extraterrestrial vampire tale) in which every stranger becomes suspect:

> I can't even begin to describe the sense of panic I feel when standing this close to other people during rush hour in the subway. Those things travel on people's clothes and bags![52]

Among the most explicitly "others" in contemporary America are undocumented immigrants, so it is not surprising that some people link bed bugs to foreigners. Although the outbreak likely emerged from within our borders, the president of the New York State Pest Management Association blamed Eastern European and South American immigrants for bringing the pests with them.[53] Others point the finger at tourists from overseas.[54] There is also speculation in the blogosphere that the bugs' resistance to DDT can be traced to overuse of this insecticide in the developing world.[55] Finally, we can be sure that emotions are running high when people express their sense of urgency by equating bed bugs with foreign terrorists and declare that New York is "under attack."[56]

Bed Bug Basics

Even those of us who haven't read Bram Stoker's novel or watched a single vampire movie have somehow absorbed the essential features of these creatures. What adolescent doesn't know that vampires come out at night to drink the blood of their victims? And how many children don't know what comes after "Good night, sleep tight . . . "?

When an insect makes the cover of the *New Yorker* and is featured in *Doonesbury*, you can be sure that it has crept into the collective imagination. And the storylines reinforce the basics (both accurate and mistaken): bed bugs feed on our blood while we sleep, they flourish in filthy places, and they are reasonable grounds for shunning people and places. The cultural zeitgeist was captured in an episode of *30 Rock* in which Alec Baldwin's snooty character has a moment of distraught catharsis and reveals to a subway car of strangers, "My name is Jack Donaghy, and I have bed bugs!" In response to the Alcoholics Anonymous–like declaration, a homeless man edges away from the Upper East Side snob.[57] The scene is funny because it touches on the truth.

Just as surely as the name Dracula conjures up a set of shared images in Western culture, the mention of bed bugs evokes a kind of real-life horror. These insects have truly infested the collective mind of America.

A CONFESSION: I'M ONE OF US

I admit that before I embarked on a recent trip to the East Coast, I checked out a website listing of hotels purportedly infested by bed bugs. Although nobody had reported the place I was staying, I took a minute when I arrived at the hotel to inspect the drawers and peek behind the headboard and picture frames. And I made sure that my suitcase was perched on one of those folding stands—not touching the wall or dresser. Paranoid or prudent? Hard to say, but I intend to stay one of us. I know how we treat them.

NOTES

1. *Bed bug* is often misspelled as *bedbug*. Entomologists separate the common name of an insect when the reference is taxonomically accurate. Hence, we refer to "dragonflies" and "ladybugs" because they are, respectively, not true flies (Diptera) or bugs (Heteroptera), while referring to "house flies" and "bed bugs" for the real deal.
2. Marshall Sella, "Bedbugs in the duvet: An infestation on the Upper East Side," New York Magazine, May 2, 2010, http://nymag.com/news/features/65733 (accessed August 20, 2012); "In search of a bedbug solution," *New York Times*, September 5, 2010, WK7.
3. Kimberly Stevens, "Sleeping with the enemy," *New York Times*, December 25, 2003, F5.
4. Meribah Knight, "The bedbug decider," *New Yorker*, February 1, 2010, http://www.newyorker.com/talk/2010/02/01/100201ta_talk_knight (accessed August 20, 2012).
5. "Bedbugs checking in at the best hotels," *New York Times*, July 26, 2001, F5.
6. Sella, "Bedbugs in the duvet."
7. Patrick Vignal and Nick Vinocur, "Paris bitten by New York–style bed bug scare," Reuters, November 16, 2010, http://www.reuters.com/article/2010/11/16/us-france-bedbugs-idUSTRE6AF4LY20101116 (accessed August 20, 2012).
8. Joel Stein, "What's so bad about bedbugs?," *Time*, October 24, 2010, http://www.time.com/time/magazine/article/0,9171,2024213,00.html (accessed August 20, 2012).
9. Cora Buckley, "Doubts rise on bedbug-sniffing dogs," *New York Times*, November 11, 2010, A1.
10. Sella, "Bedbugs in the duvet."
11. John Guenther, "Bedbugs sap Harbor Gateway woman's wallet and patience," *Daily Breeze*, December 6, 2010, http://www.dailybreeze.com/news/ci_16360616 (accessed August 20, 2012).
12. Shanna Wester, "Bed bug battle royale," State Press Magazine, October 14, 2010, http://www.statepress.com/2010/10/14/bed-bug-battle-royale-the-fight-to-kill-the-buggers (accessed August 20, 2012).

13. Online response to Sella, "Bedbugs in the duvet."
14. Wester, "Bed bug battle royale."
15. Sonya Padgett, "Fighting battle of the bedbug at home," *Las Vegas Review-Journal*, October 17, 2010, http://www.reviewjournal.com/life/fighting-battle-bedbug-home?ref=869 (accessed August 20, 2012).
16. Sella, "Bedbugs in the duvet."
17. Quoted in Knight, "The bedbug decider."
18. Quoted in Padgett, "Fighting battle of the bedbug at home."
19. Emily B. Hager, "What spreads faster than bedbugs? Stigma," *New York Times*, August 21, 2010, A1.
20. Google search, August 5, 2012. Even more hits were found with variations on the search terms (e.g., "bedbugs," "post traumatic stress disorder").
21. Susan B. Miller, *Disgust: The Gatekeeper Emotion* (Hillsdale, NJ: Analytic Press, 2004), 171.
22. Guenther, "Bedbugs sap Harbor Gateway woman's wallet and patience."
23. Petula Dvorak, "Bed bugs: Their press rivals Bristol's," *Washington Post*, September 24, 2010, http://www.washingtonpost.com/wp-dyn/content/article/2010/09/23/AR2010092306862.html (accessed August 20, 2012).
24. Jeffrey A. Weinstock, "Vampires, vampires, everywhere!," *Phi Kappa Phi Forum* 90 (Fall 2010): 4–5; Jeffrey A. Weinstock, *The Vampire Film: Undead Cinema* (New York: Wallflower, 2011).
25. Weinstock, "Vampires, vampires, everywhere!," 4.
26. Albert H. Schrut and William G. Waldron, "Psychiatric and entomological aspects of delusory parasitosis: Entomophobia, acarophobia, dermatophobia," *Journal of the American Medical Association* 186 (1963): 429–30.
27. Richard Roth, "Bed bugs block sex," video, CNN, October 24, 2010, http://www.cnn.com/video/#/video/us/2010/10/24/roth.bed.bugs.sex.cnn (accessed August 28, 2012).
28. Ibid.
29. Ibid.
30. Stein, "What's so bad about bedbugs?"
31. Online response to Sella, "Bedbugs in the duvet."
32. "Cashing in on bedbugs," video, KSDK, October 15, 2010, http://www.ksdk.com/news/local/story.aspx?storyid=221897&catid=71 (accessed August 20, 2012).
33. Online response to Sella, "Bedbugs in the duvet."
34. Weinstock, "Vampires, vampires, everywhere!," 4.
35. Shannon O'Brien, "How to prevent, eliminate bed bugs," GateHouse News Service, October 18, 2010, http://www.metrowestdailynews.com/lifestyle/gardening/x1404221294/How-to-prevent-eliminate-bed-bugs (accessed April 5, 2013).
36. Kate Murphy, "Bedbugs bad for business? Depends on the business," *New York Times*, September 8, 2010, B7.
37. Ibid.
38. Stevens, "Sleeping with the enemy."
39. Weinstock, "Vampires, vampires, everywhere!," 5.
40. Ibid.
41. Sella, "Bedbugs in the duvet."
42. Padgett, "Fighting battle of the bedbug at home."
43. Online response to Sella, "Bedbugs in the duvet."
44. Weinstock, "Vampires, vampires, everywhere!," 5.
45. "Cashing in on bedbugs."

46. Weinstock, "Vampires, vampires, everywhere!," 5.
47. Ibid.
48. Sella, "Bedbugs in the duvet."
49. Ibid.
50. Hager, "What spreads faster than bedbugs?"
51. Miller, *Disgust: The Gatekeeper Emotion*, 175.
52. Online response to Sella, "Bedbugs in the duvet."
53. Stevens, "Sleeping with the enemy."
54. "Bedbugs checking in at the best hotels."
55. Online response to Sella, "Bedbugs in the duvet."
56. Ibid.
57. Sella, "Bedbugs in the duvet."

EPILOGUE

INSECTS AS A PSYCHOLOGICAL PRECIPICE

So, what happened to me amid those grasshoppers on that summer day in 1998 such that I cannot forget—and that I strangely wanted to return? I go back to two stories, those of a monkey and a boy. Like Darwin's monkey confronting the snake but unable to leave it alone,[1] I continued to find grasshoppers enchanting while knowing that they could terrify me. And like the boy mesmerized by the two-headed fetus in the medical museum,[2] I understand being simultaneously repulsed and riveted.

Despite the characteristic quality of being attracted and repelled, I don't think the boy was disgusted or even horrified. Something even more ambiguous, complex, and profoundly human happened—and perhaps a similar kind of awareness even flickered at the edge of the monkey's consciousness. The answer to my condition lay not solely in the province of psychology but also in the territories of art and philosophy, or, more specifically, in the unsettled no man's land at the boundary of these realms.

* * *

For more than two centuries the sublime has been elicited by artists, analyzed by philosophers, and explained by psychologists—and it remains something of a mystery. A synthetic definition of the sublime is "the greatness of beauty, scale, goodness or brilliance that draws us closer for its virtue but terrifies us with its power and supremacy."[3] Immanuel Kant, arguably the greatest modern philosopher, turned his formidable mind to this vexing problem of aesthetics in 1790.[4] He systematically explored the sublime and concluded that it is essentially different from beauty. Using a series of conceptual pairings, he attempted to distinguish these two qualities of art and nature. For him, beauty and sublimity were, respectively, small and

Figure 12.1
The sublime in nature was the focus of a school of landscape painting that depicts humans as mere gnats in the face of nature's power. In this work, *Clearing Up—Coast of Sicily* (1847), by Andreas Achenbach, a barely discernible tattered flag on the rocks in the center foreground and a floating cask to the right suggest that the fury of a storm has dashed a ship against the shore just out of view (image through Wikimedia Commons).

large, short and long, day and night, youth and age, comedy and tragedy, cunning and bold, romance and friendship, wit and understanding, decorated and simple, charming and touching. While scholars tend to focus on Kant's dualism of the beautiful and the sublime, the philosopher actually formulated a triad comprising beauty, sublimity, and disgust. He recognized the Janus-faced quality of sublime and disgusting objects, with the former tinged by beauty.

Edmund Burke took up the conceptual complexity of the sublime in his seminal work, *On the Sublime and Beautiful*. Burke, a British philosopher, considered the sublime to be the strongest emotion in the human repertoire. In the grip of this feeling we are utterly transfixed, even taken outside ourselves in a kind of terrible ecstasy:

> The passion caused by the great and sublime in nature, when those causes operate most powerfully, is astonishment; and astonishment is that state of the soul, in which all its motions are suspended, with some degree of horror. In this case the mind is so entirely filled with its object, that it cannot entertain any other.[5]

While artists used violent storms, enormous vistas, and thundering waterfalls to exemplify the sublime, Burke turned to animals:

> There are many animals, who though far from being large, are yet capable of raising ideas of the sublime, because they are considered as objects of terror. As serpents and poisonous animals of almost all kinds. And to things of great dimensions, if we annex an adventitious idea of terror, they become without comparison greater.[6]

Grasshoppers held, for me, a strange middle ground between Burke's sense of terror in the presence of animate beings and inanimate entities. A single grasshopper is neither terrifying nor enormous. But as I noted in an earlier effort to make sense of my experience:

> At ten grasshoppers per square yard, one hops from underfoot with every step, and you begin to sense a continuity. If pressed into a sheet, the grasshoppers would form a continuous film over the prairie the thickness of plastic wrap. At twenty-five grasshoppers per square yard, there is riotous explosion with your every step. At forty, the chaos becomes self-perpetuating, and a rolling wave of life anticipates your next few paces. At a hundred grasshoppers per square yard, the world is transformed.[7]

Within that draw on the Wyoming prairie, I encountered an animate phenomenon of great dimensions made up of individuals that were scientifically familiar but emotionally transformed into pelting, clinging, inescapable "others."

* * *

Simon Morley, a modern scholar of art history, notes that during an experience of the sublime, "something rushes in and we are profoundly altered."[8] It is this change that fascinates and haunts me. The grasshoppers breached my barriers of objective distancing—a critical capacity for a scientist. According to Morley, the sublime is a transformative experience in which reason falters.

Kant recognized this cognitive crisis and, believing that rationality was essential to being human, tried to save us from the passion of the sublime.[9] His solution was to insert objectivity. When staring into deep space or watching breakers relentlessly pound a shoreline, we are rendered speechless—and thoughtless. We are, in Kant's estimation, simply terrified. Only when we step back and rationally contemplate our terror can we elevate our minds into the realm of the sublime.

Modern scholars have raised compelling objections to Kant's account.[10] Some have questioned whether such objectivity is psychologically possible. If we are authentically in the midst of a protosublime or terrifying experience, can we simply choose to detach from the moment? Perhaps it is a bit like that pretty-girl/old-hag optical illusion. Seeing one precludes seeing the other. To step outside our dark enchantment is to lose the emotion, leaving our rationality with nothing to contemplate.

Even if it is possible to achieve such distance, others have questioned whether it is desirable to analyze our experience of the sublime rationally. Doing so, one might argue, undermines an extraordinarily valuable opportunity in our mental life. But not all agree. Allen Carlson, one of the most influential philosophers of environmental aesthetics, contends that genuine aesthetic appreciation requires cognitive engagement. He is an advocate of positive aesthetics, the notion that all of nature is ultimately beautiful: "The positive aesthetic appreciation of previously abhorred life forms, such as insects and reptiles, seems to have followed developments in biology."[11] For Carlson, if we know the evolutionary histories and biological functions of other creatures, we will truly appreciate them for what they are. But what if something is authentically both awesome and awful?

I would argue that a deep understanding of biology provides the foundation for the negative sublime—a greatness of ugliness, scale, awfulness, or darkness that draws us closer for its malevolence but terrifies us with its power and supremacy. If science and rationality have a role to play in aesthetic appreciation, then as we learn about our psychological and evolutionary links to that which authentically repulses and enchants us, we have a basis for deep engagement. The experience of horror is biologically genuine, not something to convert into beauty through scientific machinations.

In the end, the sublime is transforming even if we think we understand its nature and origins. And when a scientist's ability to engage a subject rationally is lost, the scientist can seek either a new subject or new ways of engaging the old. I metamorphosed from offering scientific explanations to rendering humanistic accounts and artistic perspectives. Science became a necessary but not sufficient basis for my work.

* * *

It is my hope that this exploration of our psychological relationships to insects has provided you with more than a view into a self-indulgent struggle with my own infested mind. However, if you are one of the few people who don't know the strange push and pull that comes with encountering a cockroach, a hornet, a football-field-sized communal spider web, or a swarm of grasshoppers, try the following.

Go to the tallest building you can find (a cliff or chasm also works well). Approach the precipice and stop a few feet short, just at the point where you begin to feel discomfort. Take one more step and I'm betting that your heart begins to pound as you contemplate the desire to walk still closer, to draw perilously close to the edge. You feel both drawn and repelled, as if something were pulling you forward and pushing you back. Reason dictates that you won't fall if you go a bit closer. Rather, your deep, dark worry is that you will feel compelled to leap into the abyss.

You fear that you may jump, which is utterly irrational and completely believable. I know that feeling.

NOTES

1. Charles Darwin, *The Descent of Man* (1871; repr., New York: Dover, 2010), 23.
2. Stephen T. Asma, *On Monsters: An Unnatural History of Our Worst Fears* (New York: Oxford University Press, 2009), 6.
3. Jeffrey A. Lockwood, "The joy and wonder of fear and loathing," in *Trash Animals: How We Live with Nature's Filthy, Feral, Invasive, and Unwanted Species*, ed. Kelsi Nagy and Phillip David Johnson II (Minneapolis: University of Minnesota Press, 2013).
4. Immanuel Kant, *Observations on the Feeling of the Beautiful and Sublime, and Other Writings*, ed. Patrick Frierson and Paul Guyer (New York: Cambridge University Press, 2011).
5. Edmund Burke, *On the Sublime and Beautiful* (1757; repr., New York: Collier, 1965), 49.
6. Ibid.
7. Jeffrey A. Lockwood, "The joyful terror of oneness," *Wild Earth*, Spring/Summer 2004, 55.
8. Simon Morley, "Introduction: The contemporary sublime," in *The Sublime*, ed. Simon Morley (Cambridge: MIT Press, 2010), 12.
9. Kant, *Observations on the Feeling of the Beautiful and Sublime.*
10. Gordon Graham, *Philosophy of the Arts: An Introduction to Aesthetics*, 3rd ed. (New York: Routledge, 2005).
11. Allen Carlson, *Aesthetics and the Environment: The Appreciation of Nature, Art and Architecture* (New York: Routledge, 2002), 95.

INDEX